T0240110

Textile Science and Clothing Technology

Series editor

Subramanian Senthilkannan Muthu, SGS Hong Kong Limited, Hong Kong, Hong Kong

More information about this series at http://www.springer.com/series/13111

Subramanian Senthilkannan Muthu
Editor

Detox Fashion

Waste Water Treatment

 Springer

Editor
Subramanian Senthilkannan Muthu
SGS Hong Kong
Hong Kong
Hong Kong

ISSN 2197-9863   ISSN 2197-9871   (electronic)
Textile Science and Clothing Technology
ISBN 978-981-13-5228-7   ISBN 978-981-10-4780-0   (eBook)
DOI 10.1007/978-981-10-4780-0

Printed on acid-free paper

This Springer imprint is published by Springer Nature
The registered company is Springer Nature Singapore Pte Ltd.
The registered company address is: 152 Beach Road, #21-01/04 Gateway East, Singapore 189721, Singapore

This book is dedicated to:
The lotus feet of my beloved
Lord Pazhaniandavar
My beloved late Father
My beloved Mother
My beloved Wife Karpagam and
Daughters Anu and Karthika
My beloved Brother
Last but not least
To everyone working in the global textile
supply chain to make it TOXIC FREE &
SUSTAINABLE

Contents

Sustainable Waste Water Treatment Technologies

P. Senthil Kumar and A. Saravanan

Abstract Nowadays, the environmental problems associated with residual colour in industrial effluents have posed a serious threat to many environmental scientists. The effluents from the industries have liberated wide variety of pollutants which can directly introduce into the natural water sources. The industrial sector usually consumes enormous amount of water for manufacturing the sportswear, fashion and luxury brands of clothes. In that, several hazardous chemicals were added for colouring and designing purposes which contains many organic and inorganic substances, ammonia, infectious microorganisms, detergents, heavy metals, pesticides and household cleaning aids. These water pollutants are toxic to fish and other aquatic lives and it is also harmful to humans. So, there is a need for removal of toxic pollutants from the industrial effluents. The methods for controlling the water pollution can be majorly classified into three steps: (i) Primary (screening, sedimentation, homogenization, neutralization, mechanical flocculation, chemical coagulation) (ii) Secondary (aerobic and anaerobic treatment, aerated lagoons, activated sludge process, trickling filtration, oxidation ditch and pond) and (iii) Tertiary (membrane technologies, adsorption, oxidation technique, coagulation and flocculation, electrochemical processes, ion exchange method, crystallization, Evaporation). This chapter describes a critical review of the current literature available on various wastewater decolourization techniques being applied to remove the hazardous chemicals from industrial wastewater.

Keywords Industrial effluent · Hazardous chemicals · Pollution · Treatment · Colour removal

P. Senthil Kumar (✉) · A. Saravanan
Department of Chemical Engineering, SSN College of Engineering, Chennai 603110, India
e-mail: senthilchem8582@gmail.com

© Springer Nature Singapore Pte Ltd. 2018
S.S. Muthu (ed.), *Detox Fashion*, Textile Science and Clothing Technology,
DOI 10.1007/978-981-10-4780-0_1

1   Introduction

Water, air and nourishment are among a portion of the fundamental components in life. Subsequently, natural contaminations and the shrinkage of valuable items have influenced the lives of numerous (Mehta et al. 2015; Bhatnagar et al. 2011). The world comprises of a noteworthy bit (around 71%) of water however freshwater adds to just a minor portion of 2.5%. Be that as it may, more than 60 billion m^3 a time of freshwater is expected to adapt to the yearly worldwide populace development of 80 million individuals. Persistent populace development, expanding way of life, environmental change, industrialization, farming and urbanization are setting off the reduction in water asset around the world (Wu et al. 2013). The expulsion of suspended matter from water is one of the real objectives of water treatment. Lately there has been impressive enthusiasm for the improvement of utilization of regular coagulants which can be created from plants.

Water, food and energy securities are emerging as increasingly important and vital issues for India and the world. Most of the river basins in India and elsewhere are closing or closed and experiencing moderate to severe water shortages, brought on by the simultaneous effects of agricultural growth, industrialization and urbanization (Ihsanullah et al. 2016). Current and future fresh water demand could be met by enhancing water use efficiency and demand management.

Thus, wastewater/low quality water is emerging as potential source for demand management after essential treatment. An estimated 38,354 million litres per day (MLD) sewage is generated in major cities of India, but the sewage treatment capacity is only of 11,786 MLD. Similarly, only 60% of industrial wastewater, mostly large scale industries, is treated. Performance of state owned sewage treatment plants, for treating municipal waste water, and common effluent treatment plants, for treating effluent from small scale industries, is also not complying with prescribed standards.

According to World Mapper Project (2007), 990 billion m^3 of water utilized for domestic and industrial purpose worldwide each year and then this freshwater is transformed into wastewater. This wastewater mostly comprises of hazardous chemical which are persistent in nature. They can gradually accumulate in the food chain, in turn, can cause long-term, irreversible damage to people like cancer, delayed nervous damage, malformation in urban children, mutagenic changes, neurological disorders etc. (Qu et al. 2013). And they also have serious impact upon environment such as eutrophication or oxygen depletion in lakes and rivers. Therefore, many Environmental laws were enacted and their enforcement also made stricter.

The textile manufacturing utilizes an assortment of chemicals, contingent upon the way of the crude material and final result. Some of these chemicals are diverse compounds, cleansers, colours, acids, soft drinks what's more, salts. Material finishing sector utilizes a lot of water, basically as a result of colouring and cleaning/washing operations. Clearly the wastewater gushing created from these units contains extensive measures of dangerous toxins (Paul et al. 2012). On the off chance that these wastewaters are released into nature they will cause genuine and unsafe effect not just on underground and surface water bodies and land in the encompassing region additionally will adversely affect the sea-going biological framework. Due to utilization of colours and chemicals, effluents are dull in shading, which builds the turbidity of water body.

The qualities of industrial effluents fluctuate and they basically rely upon the kind of material made furthermore, the chemicals utilized. The industrial wastewater gushing contains high measures of operators harming nature and human wellbeing and also it includes suspended and broke down solids, natural oxygen request (BOD), compound oxygen request (COD), chemicals, contain follow metals like Cr, As, Cu and Zn (Qu et al. 2013; Rao et al. 2006).

At whatever point great quality water is rare, water of marginal quality should be considered for use in farming. In spite of the fact that there is no widespread meaning of 'marginal quality' water, for all functional purposes it can be characterized as water that has certain attributes which can possibly bring about issues when it is utilized for an expected reason. For instance, brackish water is marginal quality water for farming use due to its high broke up salt substance, and metropolitan wastewater is negligible quality water as a result of the related wellbeing perils. From the perspective of water system, utilization of "negligible" quality water requires more perplexing administration practices and more stringent observing techniques than when great quality water is utilized. Development of urban populations and expanded scope of domestic water supply and sewerage offer ascent to more prominent amounts of metropolitan wastewater. With the current emphasis on ecological health and water contamination issues, there is an expanding attention to the need to discard these wastewaters securely and gainfully.

Industrial wastewater is principally included water (99.9%) together with generally little concentrations of suspended and dissolved organic and inorganic solids. Among the organic substances shown in sewage are starches, lignin, fats, cleansers, manufactured cleansers, proteins and their disintegration items, and in addition different common and engineered natural chemicals from the process industries. Presence of sulfur, naphthol, vat colours, nitrates, acidic corrosive, cleansers, chromium mixes furthermore, substantial metals like copper, arsenic, lead, cadmium, mercury, nickel, and cobalt and certain assistant chemicals all on the whole make the gushing profoundly harmful. Other harmful chemicals exhibit in the water might be formaldehyde based colour settling operators, hydro carbon based conditioners and non bio degradable colouring chemicals. The process gushing is likewise regularly of a high temperature and pH, both of which are to a great degree harming. Scouring, dyeing, printing and finishing are processes generating the majority of industrial wastewater (many rinsing sequences after each step). Significant variation of ecological parameters of industrial waste water effluents was shown in Table 1.

Be that as it may, auxiliary chemicals and unintended debasement items may likewise be available in the materials and cause harmful impacts on human health and the earth, yet these sorts of substances are not secured by screening concentrate because of the restrictions.

Table 1 Significant variation of ecological parameters of industrial waste water effluents

S.No.	Constituents	Concentration
1	pH	2–13
2	COD	10–61,900 (g/m^{-3})
3	TSS	5–7630 (g/m^{-3})
4	Turbidity	1–200 (NTU)
5	Conductivity	0.2–115.2 (mS/cm)
6	Color-absorbance average	0.001–218.8
7	TDS	5–1170 (g/m^{-3})
8	BOD	5–770 (g/m^{-3})
9	Nitrogen	20–150 (g/m^{-3})
10	Phosphorus	2–25 (g/m^{-3})

2 Wastewater Characteristics

From the textile industry, it has been assessed that 90% of wastewater or 90,000 tons that goes untreated, while just 10% is reused. Furthermore, 10–15% of a overall 700,000 tons of dye production is disposed of as waste. For the wastewater qualities, it has been found that most of the squanders from the colouring business is from the different colouring forms (batch and continuous), soluble planning, and likewise the constituents from colouring, for example, salts included inside a portion of the chemical to create colours for the different procedures in the industry. In particular, the attributes of emanating included chemical oxygen demand (COD), high biochemical oxygen demand (BOD) where an expected focuses incorporate the most astounding BOD was in fleece scouring, complex handling, and covering completing (2270, 420, and 440 mg/L individually). In addition to that, chemical oxygen demand (COD) has been seen as 2 and 13 times more than the BOD focuses—12 as COD focus was at 7030 mg/L when contrasted with 2270 mg/L for fleece scouring wastewater creation. Different segments that have been seen incorporate the nearness of total suspended solids (TSS), where fleece scouring revealed 3310 mg/L and oil and grease (O&G). When one considers colouring wastewater particularly, the wastewater comprises of metals, salts comprising of magnesium chloride and potassium chloride (2000–3000 ppm), surfactant, toxics, shading, BOD, COD, sulphide. At last, toxicity of the wastewater has been seen fluctuating in view of the nearness of constituents. From a trial of 75 material factories, it was watched that 38 or more than half of the 75 material plants had no lethality, while around 9% had harmful segments. Potential wellsprings of lethality incorporate salts from the colouring status, surfactants, metals from the colours, organics. Particularly with colours, it was found that 63% of 46 tried business colours had a poisonous quality range utilizing the deadly fixation (LC50) esteem, or the focus required to execute half of a given populace, measured more noteworthy than 180 ppm or having little danger, while just 2.2% were viewed as harmful. Wastewater quality can be defined by its physical, chemical and biological constituents present in wastewater.

2.1 Physical Characteristics

Wastewater temperature is a key parameter because it affects the chemical and biological reactions of microorganism. High temperature can increase undesirable plank tonic species and fungi. Other various parameters such as pH, conductivity, saturation level of gases and various form of alkalinity etc., depends on temperature (Raeesossadati et al. 2014).

The colour usually represents the age of wastewater. The industrial wastewater appearance depends mostly upon the nature of the product manufactured. Odours are released from wastewater due to dissolved impurities, organic nature caused by living and decaying aquatic organisms, and accumulation of gases. The total solids contents are denoted by various types of dissolved and suspended material remained as residues in wastewater (Metcalf and Eddy 1987).

2.2 Chemical Characteristics

Organic materials are normally composed of carbon, hydrogen, and oxygen. Presence of ammonia and nitrogenous matter in the wastewater can be accepted as the chemical evidence of organic pollution (Sahu et al. 2013). The principal components of wastewater are proteins, carbohydrates, lipids, oils and urea and small quantities of several synthetic organic chemicals (Jiang et al. 2011). The common inorganic compounds in wastewater are chloride, hydrogen, iron, Table 1. Typical range of BOD and TSS load for industrial and domestic wastewater (Source: Industrial wastewater treatment plants self-monitoring manual, Chap. "Review of Utilization Plant-Based Coagulants as Alternatives to Textile Wastewater Treatment", 2002) 4 nitrogen, phosphorus, sulfur and trace amounts of heavy metals (Choi and Lee 2015).

2.3 Biological Characteristics

Naturally, wastewater contains large amounts of macro and microscopic organisms. Based on the quantity and potential of biological organism present in wastewater, the effectiveness of the treatment facilities can be determined.

3 Substances of Potential Hazard to Human Health

The textile related substances for example, auxiliary chemicals and impurities/degradation products, which can be of potential hazard to the human health. The concentration of such substances is lower in the last material article than the centralization of practical chemicals and therefore they were rejected from the extent of the screening study. Around 40% of all inclusive utilized colorants contain naturally bound chlorine a known cancer-causing agent. All the natural materials exhibit in the wastewater from a material industry are of extraordinary worry in water treatment since they respond with numerous disinfectants particularly chlorine. Chemicals dissipate into the air we inhale or are consumed through our skin and appear as hypersensitive responses and may cause harm to foetus. Azo colours are additionally profoundly dangerous to the biological system and mutagens, which means they can have intense to endless impacts upon life forms, contingent upon introduction time and dye concentration. For instance, dye effluent has been connected to development decrease, neurosensory harm, metabolic anxiety and passing in fish, and development and profitability in plants. Contamination along these lines restrains downstream human water utilizes, for example, entertainment, drinking, angling and water system.

The exercises to treat risky wastes can run from lawful preclusion to cost sparing reusing of chemicals. Contingent upon the sort of item and treatment, these means can appear outrageous fluctuation. In this way it is basic to clean the wastewater before release keeping in mind the end goal to ensure our common habitat from unsafe impact of the profluent.

Effluents treatment plants are the most broadly acknowledged methodologies towards accomplishing natural wellbeing. Be that as it may, shockingly, no single treatment technique is appropriate or generally adoptable for any sort of profluent treatment. In this manner, the treatment of waste stream is done by different strategies, which incorporate physical, substance and natural treatment contingent upon contamination stack. Our point is to embrace advancements giving least or zero natural contamination.

4 Wastewater Treatment Strategies

Water reuse and supportability will keep on being critical objectives for ecological contamination aversion/lessening hones in the material business. The industries will proceed to pick and use water treatment arrangements to lessen its working expenses, as well as to diminish its water impression and decline the biological effect from its wastewater release and solids sludge generation on the surrounding ecosystem. Wastewater treatment process optimization will keep on being a point of convergence for industrial organizations as the expenses of wastewater disposal and freshwater utilization keep on escalating due water shortage issues.

4.1 Unit Operations for Wastewater Treatment

There are a great deal of pollutants and squanders in the wastewater, for example, supplements, inorganic salts, pathogens, coarse solids and so forth, which are truly perilous for environment and human, for evacuating these poisons, diverse procedures have been exposed. There are particular procedures and unit operations in sewage/wastewater treatment, the essential objective of these procedures is to diminish the contamination of the water dirtying beginning stage until the end of the treatment procedure which can be transfer or reusal and these diminishment forms can be substance, physical or natural. Unit operations for water treatment can be classified as physical, chemical and biological (Lettinga and Pol 2008; Oreopoulou and Russ 2006).

4.1.1 Physical Unit Operation

Physical unit operation are some treatment strategies which wash down the wastewater by utilizing the physical powers, for example, flocculating, floatation, blending, filtration, screening and gas exchange.

4.1.2 Chemical Unit Processes

Chemical unit procedures are the methods that cause responses in wastewater parts such procedures are utilized while the physical and natural procedures are in activity (Harremoes 2002). There are a considerable amount of various compound procedures, for example, precipitation, coagulation, neutralization and stabilization, ion exchange, oxidation and advanced oxidation that might be added to sewage water amid the cleaning system (Lettinga and Pol 2008).

4.1.3 Biological Unit Processes

Biological unit processes is the systems that separate the oil/oil, Suspended solids, natural matter, nitrogen and phosphorus by microorganisms which develop actually in an organic reactor. The microscopic organisms expends the carbon-based material in the sewers, moreover the essential objective of this treatment is to diminish the organic components in wastewater (Rosen and Lofqvist 1998). The benefits of using a biological method are as follows:

- A biological method is natural and does not require the use of unsustainable chemicals.
- Once the system has been set up, the running costs are lower.
- Instead of having a process to remove either P or N, the biological process can remove both P and N whilst simultaneously sequestering C, reducing other trace nutrients and providing oxygen for improved BOD reduction through an algae/bacteria co-symbiotic relationship.
- A biomass by-product is created that can be used as an energy source or fertiliser.

- The water utilities are familiar with bio-processes.
- This method is more appropriate when the wastewater contains BOD/COD ratio greater than 0.25.

An above stated treatment technique such as physicochemical treatment and biological treatment are mostly used for wastewater. Since none of the techniques are capable to treat wastewater completely, physicochemical methods are coupled with biological methods. Because physicochemical methods are restricted due to high cost, alteration in the influent characteristics and also they are suitable only at low pollution load. Whereas biological treatments are capable of attacking raw effluent directly with less sludge left over. But, a lot of research works are needed for strain selection or its consortia for optimization and to improve performance of the treatment of wastewater.

All things considered, treatment made by utilizing organic has been censured due to its impediment of biodegradability by microorganisms, especially due to the xenobiotic parts that can be found within biological treatment processes. Notwithstanding, it has been watched that the utilization of routine organic techniques is unequipped for expelling colour from wastewater because of the nearness of numerous organic contaminants. This is because of the way that natural mixes inside colours are exceptionally instable and have high imperviousness to natural matter deterioration. The chemical components of the dye are exceptionally troublesome for the microorganisms to have the capacity to debase down which is the reason it is troublesome for traditional organic treatment to be utilized for the treatment of colour wastewater (Fig. 1).

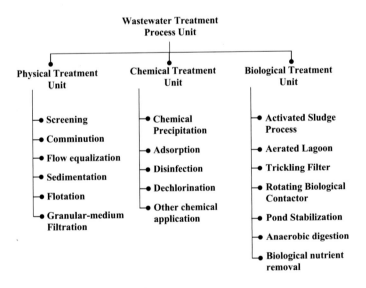

Fig. 1 Common wastewater treatment processes

4.2 Phases of Water Treatment Methodologies

Water shortage and quantity of grey/wastewater is increasing with rapid expansion of industries and domestic water supply. Thus it is essential to purify and reuse the wastewater to reduce the burden. In view of the aforesaid problems, recent attention has been focused on the development of more effective, lower-cost, robust methods for wastewater treatment (WWT), without further stressing the environment or endangering human health by the treatment itself.

Physical treatment acts as basic processing step for most wastewater treatment processes. In this treatment, the physical units are used for the treatment which follows mechanical forces. Likewise, in chemical process some chemical reactions are performed to bring some alterations in the treatment. This chemical process is usually accompanied either with the physical or biological processes to enhance the treatment process (George et al. 2015). In biological process, the biological organisms are responsible for the transform of dissolved and particulate biodegradable constituents into acceptable end products.

Different treatment techniques have been employed for wastewater treatment. Preliminary, Primary, secondary and tertiary treatment processes are utilized to expel contaminants from waste-water. In request to accomplish diverse levels of contaminant evacuation, individual waste-water treatment methodology are joined into an assortment of frameworks such as filtration, electro flotation, electrocoagulation, aerated lagoons, trickling filters, aerobic activated sludge, electrolysis, reverse osmosis, ion exchange, adsorption and advanced oxidation process.

4.2.1 Preliminary Treatment

The goal is to expel the expansive materials like coarse solids which are all considered a significant part of the time found in wastewater. Besides, it isolates the gliding materials which are being conveyed by water stream. Preliminary treatment methodology as a rule contains coarseness expulsion, coarse screening and comminution of expansive articles. Also this treatment helps in evacuating the greases and oils. This procedure diminishes the wastewaters BOD, by roughly 15–30% and the gadgets which are being utilized amid this treatment are Grit chamber and Comminutor (Kawamura 2000).

4.2.2 Primary Treatment

The initial step is the evacuation of suspended solids, extreme amounts of oil, oil and dirty materials. The effluent is initially screened for coarse suspended materials, for example, yarns, build up, bits of textures and clothes utilizing bar screens and fine screens. The screened effluents then experience settling for the evacuation of the suspended particles. The drifting particles are evacuated by mechanical scratching frameworks (Qasim 1998).

The goal is to expel strong parts of wastewater by sedimentation, these segments can be organic components, for example, phosphorus, nitrogen, and metals associated with strong parts. In addition to that, colloidal and broke up components will remain and not be influenced. The waste from essential sedimentation units is known as essential gushing furthermore, the squanders which have been delivered by this procedure is called primary effluent.

4.2.3 Secondary Treatment

This treatment is utilized after the secondary treatment which finishes the purging procedure through diminishing the measures of remaining organic components and solid particles; likewise biodegradable evacuation and colloidal natural matter utilized high-impact organic in secondary treatment processes (Tilley 2011).

Organic matter is a wellspring of vitality and supplements for oxygen consuming microscopic organisms. They oxidize natural matter to from CO_2 and water corrupt nitrogenous natural matter into ammonia. Circulated air through Aerated lagoons, trickling filter and activated sludge frameworks are among the oxygen consuming framework utilized as a part of the secondary treatment. Anaerobic treatment is principally used to balance out the sludge in this manner created.

The fascination in this secondary technique is because of its ability of microorganisms of having the capacity to mineralize organics actually, i.e. giving a common advantage to both the treatment plants and furthermore the living beings itself—waste is utilized as a food source, while it is changed over into a shape reasonable for release with little or minimum cost. All in all, the significant contrast in treatment forms rely upon regardless of whether sub-atomic natural is available as in oxygen consuming treatment, or truant as in anaerobic ().

There are two noteworthy orders of oxygen consuming treatment—trickling filter and activated sludge.

(i) Trickling Filters

Trickling filters are another normal technique for secondary treatment for the most of the part working under vigorous conditions. The effluent for the primary treatment is streamed or splashed over the channel. The channel normally comprises of a rectangular or round bed of coal, rock, poly Vinyl chloride (PVC), broken stones or engineered tars.

Trickling filter gives media by which supports microbial development, where cases incorporate pulverized stone, slack, or other inorganic materials. Whenever water enters from the highest point of the framework, it reaches the media, whereby starting development of microorganisms and expulsion of natural materials. In view of the structure of the trickling filter, it is essential that a trickling filter requires the utilization of distribution or the reusing of the arrival of treated wastewater once again into the channel with the end goal of keeping up adequate air circulation and furthermore holds dampness of the media without compromising the loss of microorganisms. One can state that evacuation proficiency is coordinated with the BOD stacking rate. Inside the material business, 10–90% of treatment can be accomplished by a trickling filter.

A coagulated film made up of microorganisms, is framed on the surface of the filter medium. These creatures help in the oxidation of natural matter in the effluent to carbon dioxide and water. Trickling filters don't require enormous space, and in this way are worthwhile as looked at to the circulated air through tidal ponds. Nonetheless, their disservice is the high capital cost and scent discharge.

(ii) Aerobic Activated Sludge

Aerobic activated sludge process is the most ordinary one. Activated sludge is the utilization of a suspended development that gives imply contact amongst microorganisms and natural constituents. It includes a standard air circulation of the effluent inside a tank permitting the aerobic bacteria to metabolize the soluble and suspended organic matters. In a perfect world, the objective is to decrease the biochemical oxygen request (BOD) that is produced from the generation of waste from different enterprises. The EPA has expressed that inside the material industry enacted slop can be accomplished at efficiencies as high as 95%. For the reason of creating nitrification or an expansion in cherished contact, one can consider the utilization of broadened air circulation. Broadened air circulation is past the routine 6–8 h muck maintenance time, where it can be reached out up to three days. Points of interest of this extra time builds the digestion of natural mixes in the reactor where over 75% of segments can be successfully utilized and diminish the measure of waste era.

A piece of the natural matter is oxidized into CO_2 and the rest are incorporated into new microbial cells. The effluent and the sludge produced from this process are isolated utilizing sedimentation; a portion of the sludge is come back to the tank as a wellspring of microorganisms. A BOD expulsion proficiency of 90–95% can be accomplished from this procedure however is tedious. Sludge formed as a result of primary and secondary treatment processes represent a noteworthy transfer issue. They cause ecological issues when discharged untreated as they comprise of microorganisms and natural substances. Treatment of sludge is completed both, aerobically and anaerobically by microscopic organisms. Oxygen consuming treatment includes the presence of air and aerobic bacteria which change over the slime into carbon dioxide biomass and water. For compelling treatment, it has been expressed that if treatment has a N to BOD proportion of 3–4 lb N/100 lb of BOD treatment and broke down oxygen fixation is kept up around zero inside aeration basins.

(iii) Aerated Lagoons

Aerated lagoons are one of the ordinarily utilized organic treatment processes. This comprises of huge holding tank lined elastic or polythene and the emanating from the primary treatment is circulated air through for around 2–6 days and the framed slime is expelled. The BOD expulsion effectiveness is up to 99% and the phosphorous expulsion is 15–25%. The nitrification of ammonia is additionally found to happen in circulated air through tidal ponds. The significant disservice of this strategy is the huge measure of space it involves and the danger of bacterial sullying in the tidal ponds. The expulsion of substantial metals, for example, Pb, Cd, Zn, Cu and Cr by utilizing duckweed and green growth lakes as a cleaning step was found to not be exceptionally reasonable.

(iv) **Anaerobic Treatment**

As expressed before, the distinctions in organic treatment is dependent upon the presence or absence of molecular oxygen. With the absence of molecular oxygen, anaerobic conditions endure. Inside this specific treatment condition hold on, the presence of microorganisms will utilize elective wellsprings of oxygen, (for example, sulfates, nitrates for instance) and change over organics into natural acids and alcohols. Promote change of these constituents would occur into methane and carbon dioxide. Anaerobic treatment can be considered in many occurrences ideal over oxygen consuming conditions because of its capacity to decrease waste and deliver an element that can be utilized as a significant asset.

Anaerobic lagoons are one of two noteworthy sorts of anaerobic conditions. An anaerobic lagoon is a misery that comprises of a profundity of 10–17 ft, a BOD stacking rate in the vicinity of 15 and 20 lb BOD/1000 ft^3, and a long sludge retention time. Wastewater normally spills out of the base of the tidal pond to guarantee the correct passage and maintainability of nourishment for anaerobes. An option framework known as an anaerobic contact framework comprises of an equalizing tank, digester with gear for blending, gas stripping utilizing air or vacuum, and clarifiers.

Anaerobic absorption starts inside the digester at respectably high temperature (95–100 °F) with BOD loadings going from 0.15 to 0.2 lb/ft^3 for a time of 3–12 h. This procedure takes after passage into a gas stripper, settling inside the clarifier, and after that reused back inside the framework. Anaerobic contact framework is equipped for accomplishing a 90–97% expulsion rate of BOD and suspended solids. Anaerobic treatment processes are fit for delivering higher treatment than aerobic. Shading evacuations have been reported to being 65% to a most extreme in the vicinity of 80 and 90% (Wijetunga et al. 2010; Somasiri et al. 2008).

The mix of anaerobic and aerobic treatment has discovered more achievement when endeavouring to treat colour wastewater. One reason why is on the grounds that anaerobic- oxygen consuming treatment has the capability of having the capacity to expel toxins while delivering methane with the end goal of being utilized for vitality. Different literatures have watched more achievement when utilizing this specific application than treatment utilizing essentially anaerobic and aerobic treatment by them.

4.2.4 Tertiary Treatment

The target is to evacuate the particular wastewater constitutes which cannot be expelled by auxiliary treatment including lethal substances, organic components and solid particles. Tertiary evacuation utilizes the flood of a waterway for reusing or groundwater restoration (Donald and Rowe 1995). Tertiary treatment aims at effluent polishing before being discharged or reused and can consist the removal of nutrients (mainly nitrogen and phosphorous), toxic compounds, residual suspended matter, or microorganisms (disinfection with chlorine, ozone, ultraviolet radiation or others). Nevertheless, this third stage/level is rarely employed in low-income countries. Tertiary treatment process can include membrane filtration (micro, nano, ultra and reverse osmosis), infiltration/percolation, activated carbon, disinfection (chlorination, ozone, UV) (Geise et al. 2010). Finally, water reclamation refers to the treatment of wastewater to make it suitable for beneficial use with no or minimal risk.

(i) **Ultrafiltration**

Ultra filtration membranes can be produced using both natural (polymer) and inorganic materials. Ultra filtration is a low-weight film process used to separate high atomic weight mixes, colloidal materials, organic and inorganic polymeric molecules from a feed stream. There are a few polymers and different materials utilized for the maker of UF layer. The decision of a given polymer as a film material depends on certain properties, for example, sub-atomic weight, chain adaptability, and chain connection. The structure of UF layer can be symmetric or asymmetric. The thickness of symmetric membrane (permeable or nonporous) is range from 10 to 200 μm. The imperviousness to mass exchange is dictated by the aggregate film thickness. A reduction in layer thickness comes about in an expanded penetration rate. Ultra filtration layers have an uneven structure, which comprise of exceptionally thick top layer or skin with thickness of 0.1–0.5 μm upheld by a permeable sub layer with a thickness of around 50–150 μm (Al-Bastaki and Banat 2004).

Ultra filtration has larger pores than nanofiltration and reverse osmosis. Low molecular-weight organics and ions such as sodium, calcium, magnesium chloride, and sulfate are not removed. Ultra filtration utilizing adjusted poly vinylidene fluoride film which has 60% of styrene-acrylonitrile in substance and 40% poly vinylidene fluoride with a permeable top layer and a sub-layer with various pores. The shading expulsion and COD diminishment is direct joined by lessened layer fouling for partition and decontamination of colour arrangements. The treatment by ultrafiltration and/or nano filtration invalidate a portion of the impediments of the layer procedure, for example, fouling, pore blocking and cake arrangement and empowers water reuse.

(ii) **Nanofiltration**

Nanofiltration (NF) is the most recently developed pressure-driven membrane process for liquid-phase separations. The properties of NF films lie between those of non-permeable RO films (where transport is represented by solution-diffusion mechanism) and permeable ultrafiltration (UF) films (where division is generally thought to be because of size rejection and, at times, charge impacts). NF is one of the promising innovations for the treatment of characteristic natural matter and inorganic pollutants in surface water. Since the surface water has low osmotic weight, a low-weight operation of NF is conceivable. In the NF of surface waters, natural organic compounds, which have generally extensive particles contrasted with layer pore size, could be expelled by sieving instrument, though the inorganic salts by the charge impact of the films and particles.

NF has favourable circumstances of lower working weight contrasted with RO, and higher organic dismissal contrasted with UF. For the colloids and extensive atoms, physical sieving would be the prevailing dismissal component though for the particles and lower atomic weight substances, arrangement dispersion component and charge impact of layer assume the significant part in separation process. The nano filtration strategies accomplish a sharp diminishment in COD alongside the colours expulsion from the pervade. The cross stream nanofiltration with the assistance of a thin film composite polysulfone film working at low weights, generally high fluxes are gotten, with a normal colour dismissal of 98% and NaCl dismissals of fewer than 14%. Accordingly, a high calibre of reusable water is recovered. NF can be utilized as a part of the tertiary wastewater particularly to expel continuing organic toxins. The end goal to enhance the filtration flux of NF and expand the operation of NF without broad natural fouling, viable pretreatment is vital. As of late, high rate flocculation and magnetic ion exchange resin have been attempted to evacuate hydrophobic and hydrophilic organics individually. This can incredibly diminish the organic fouling on the NF layers.

(iii) Reverse Osmosis

Reverse osmosis technology has numerous applications in the wastewater treatment. This RO focus will have an abnormal state of natural fixation. As the penetrate acquired is exceptionally immaculate it is broadly by and by as of late. The cationic mixes i.e. metals are expelled by giving a high hydrostatic weight over the layer. Polyamide when utilized as a skin material for RO layer brought about entire expulsion of Cu(II) and Ni(II) metals. It likewise demonstrates that the dismissal rate of metal particles increments with the expansion in the transmembranic pressure. Reverse osmosis is a procedure that industry uses to clean water, regardless of whether for modern process applications or to change over bitter water, to tidy up wastewater or to recover salts from mechanical processes. Reverse osmosis won't expel all contaminants from water as disintegrated gasses, for example, dissolved oxygen and carbon dioxide not being expelled. Be that as it may, reverse osmosis can be exceptionally viable at evacuating different items, for example, trihalomethanes (THM's), a few pesticides, solvents and other unpredictable natural mixes (VOC's).

Reverse osmosis membranes is a common membrane materials include polyamide thin film composites (TFC), cellulose acetate (CA) and cellulose triacetate (CTA) with the membrane material being spiral wound around a tube, or hollow fibres bundled together. Hollow fibre membranes have a greater surface area and hence capacity but are more easily blocked than spiral wound membranes. RO membranes are rated for their ability to reject compounds from contaminated water. A rejection rate (% rejection) is calculated for each specific ion or contaminant as well as for reduction of total dissolved solids (TDS). TFC membranes have superior strength and durability as well as higher rejection rates than CA/CTA membranes. They also are more resistant to microbial attack, high pH and high TDS. CA/CTA's have a better ability to tolerate chlorine. Sulphonated polysulphone membranes (SPS) are chlorine tolerant and can withstand higher pH's and are best used where the feed water is soft and high pH or where high nitrates are of concern.

The performance of a system depends on factors such as membrane type, flow control, feed water quality, temperature and pressure. Also only part of the water entering the unit is useable, this is called the % recovery. This is affected by the factors listed above. For example the amount of treated water produced can decrease by about 1–2% for every 1 °C below the optimum temperature. Systems must be well maintained to ensure good performance without any fouling and maximising the output of water. Biocides may be needed and the choice of biocide would depend on the membrane type, alternatively other filters may be required to remove chlorine from water to protect the life of the membranes. To this end a good treatment regime is needed and knowledge of the specific foulants so the optimum cleaning and maintenance chemicals can be chosen.

(iv) **Electrocoagulation**

Contradictory to electro coagulation is the utilization of coagulation-flocculation. This technique includes the utilization of different coagulants, customarily alum (aluminum sulfate), ferric chloride ($FeCl_3$), or ferrous sulfate (Fe_2SO_4) which can be extremely costly contingent upon the volume of water treated (Bratby 2006). After application the coagulant, charge of the particulates agglomerate and settle at the base of the tank. Substance coagulation/flocculation is worried with the pH, blending and time (Verma et al. 2012; Bidhendi et al. 2007; Wong et al. 2007).

Electro coagulation with aluminium or the hybrid Al/Fe electrodes is reasonable for water generation or wastewater treatment. This procedure has been turned out to be extremely powerful in expelling contaminants from water and is described by lessened sludge generation, no necessity for synthetic utilize and simplicity of operation. The Electro coagulation method has been seen to be more viable for the expulsion of COD than the traditional coagulation and sedimentation forms. Solvent metal cathodes like Al and Fe were observed to be exceptionally powerful in contrast with insoluble terminals, for example, carbon (C), and titanium (Ti). Al and Fe particles support to the coagulation of colloidal particles. In this strategy for treating dirtied emanating, conciliatory anodes (Al and Fe) erode to discharge dynamic coagulant antecedents into the wastewater. These particles deliver insoluble metallic hydroxides of Al and Fe which can expel poisons by surface complexation or electrostatic fascination.

Uses of plant-based coagulants are comparable to their substance partners as far as effectiveness. This speaks to critical advance in reasonable natural innovation as they are renewable assets and their application is straightforwardly identified with the change of personal satisfaction in the biological community.

Electro flotation is likewise a technique for isolating substances in which electrically created small rises of hydrogen and oxygen gas connect with poison particles making them to coagulate and glide on the surface of water body. The electro flotation innovation is viable in expelling colloidal particles, oil and grease, and in addition natural poisons with preferred execution over either broke up air flotation, sedimentation, impeller flotation. Electro deposition is viable in recuperation of heavy metals from wastewater streams.

It is extremely hard to discover routine utilization of compound coagulation/flocculation systems the same number of have turn towards the utilization of joining treatment techniques for better upgraded treatment. One of the conceivable issues with utilizing synthetic coagulation/flocculation alone is the trouble of being ready to lessen solvency enough for segments to have the capacity to shape flocculants to be expelled from the wastewater.

(v) Electrolysis

The "Electrolysis" actually intends to break substances separated by utilizing power. The procedure happens in an electrolyte, a watery or a salt dissolving arrangement that gives a probability to exchange the particles between two terminals. At the point when an electrical current is connected, the positive particles move to the cathode while the negative particles move to the anode. At the terminals, the cations are lessened and the anions will be oxidized. Earth arranged electrochemistry is increasingly requested contamination decrease of wastewater and recovering the necessity of release or reasonable breaking point of wastewater. Under these conditions an electrochemical treatment is a rising innovation with numerous applications in which an assortment of undesirable broke up dangerous chemicals and microorganisms can be adequately expelled from wastewater. The principle forms happening amid electrolysis are electrolytic responses at the surface of electrodes, arrangement of coagulants in watery stage, adsorption of solvent or colloidal poisons on coagulants, and expulsion by sedimentation and floatation.

Accordingly of electro synthetic responses, the dissolved metal particles join with finely scattered particles in the arrangements, shaping heavier metal particles that encourage and can be evacuated later. One of the disserves is that a high contact time is required between the cathode and the effluent.

Factors affecting electrolytic treatment and process performance

The control, operation and substance cooperation of the electrolytic framework influence the execution and dependability of electrolytic treatment innovation. Adding to multifaceted nature and the reasonable contaminant evacuation systems and their connections with the reactor plan, current thickness, terminal sort and working time impact the electrolysis.

(a) Reactor Design

The reactor design influences operational parameters including bubble way, flotation viability, floc-development, liquid stream administration and blending/settling qualities. It is imperative to outline the reactor for a particular procedure and the reactors for energy transformation and electrochemical union will have diverse drivers to those utilized as a part of the obliteration of electrolyte-based contaminants. The type of the reactants and items; and the method of operation (bunch or ceaseless) are additionally the vital plan components.

(b) **Applied Current Density**

Applied current density assumes critical part in electrolytic treatment as it is the main operational parameter that can be controlled straightforwardly. In this framework cathode separating is settled and current is persistently provided. After destabilization of the colloidal suspension, powerful accumulation requires sufficient contact present and more coagulant (Al) accessible per unit of time. The living arrangement time is diminished in the reactor, decreasing the likelihood of crash and attachment amongst contamination and coagulant. Current thickness straight-forwardly decides both coagulant measurements and bubble generation rate; and firmly impacts both arrangement blending and mass-exchange at the electrodes.

(c) **Electrode Type and Arrangement**

The wastewater to be dealt with is gone through the electrolytic reactor with anodes and was subjected to coagulation and flotation by creating the particles framing the cathodes. These particles skimming on the surface of wastewater in the wake of being caught by hydrogen gas air pockets are produced at cathode surfaces. The anode associations in an electrolytic reactor are monopolar and bipolar.

(d) **Operation Cost**

The way toward assessing and choosing proper wastewater treatment innovation ordinarily starts with a specialized possibility concentrate that relies on upon the way of the application. The most fundamental perspective that should be considered to assess the capital wanders of a treatment. The operational cost includes expenses of chemicals, anodes and vitality utilizations and in addition works upkeep, sludge dewatering and transfer, and settled expenses.

(vi) **Ion Exchange**

Ion exchange is a physical separation process in which the particles in arrangement are exchanged to a solid lattice. Ion Exchange can be used in wastewater treatment plants to swap one ion for another for the purpose of demineralization. Ion exchange resins can be characterized on the premise of useful gatherings as anion exchange resin, cationic exchange resin and chelating exchange resin (Kumar et al. 2017). The materials can be further broken down to individual grouping depending on whether it is a strong base cation, weak acid cation, strong base anion, or weak base anion.

Ion-exchange resins are small, porous beads that are negatively or positively charged, allowing them to grab onto ions (contaminants in the water) that are attracted to that charge. These resins are solvents (insoluble in water), and they range in diameter from 0.3 to 1.5 mm. Resin is placed in a vessel, usually called a column, and submerged in water where it forms a layer on the bottom called a bed. The bed absorbs water and swells when it is first immersed. Immersion conditions the resin. When the resin is fully conditioned, the beads contain 50–70% water.

Wastewater is passed through the resin columns while the resin bed is gently agitated. The agitation allows the water to flow uniformly around the resin beads. The agitation actually increases the amount of surface area that comes in contact with a wastewater, which increases the likelihood that the porous openings will come into contact with ions. Imagine the resin bead as a ball covered in holes. As the ball rolls and bounces in the wastewater, its holes become exposed to the particles suspended in the water. Due to the charge, if the ions come into contact with the resin, they will be attracted to it and become trapped in the pores

Ion exchange resins having particular metal take-up limit are being utilized during ion exchange process. The resins might be either manufactured or characteristic. Manufactured resins are significantly utilized due to their preeminent metal expulsion limits from the watery arrangement. Naturally occurring silicate minerals called as zeolites assume an imperative part in ion exchange process.

For instance, in a common demineralizer, influent water which passes through a cation exchange resin will be stripped off from it metallic cations salt to become acids whereby loss of the ions will be replaced with a similar corresponding amount of hydrogen ions. The resultant acids will then be removed through another alkaline regenerated anion exchange resins in which this time round, the anions present in the wastewater will be substituted with equivalent amount of hydroxides. As the capacity of the bed always has a certain fixed limit, eventually the resin will become exhausted and thus has to undergo regeneration process. The cation exchange resin is regenerated using either hydrochloric or sulphuric acid, producing waste brine in the process while the anion resin will be regenerated with sodium hydroxide. The benefits of using a ion exchange resin are as follows:

- No sludge generated. If the wastewater being treated is from an electroplating operation, for example, sludge is considered F006 hazardous and can be very expensive to haul off.
- Less labour intensive than chemical treatment.
- Columns ship easily and are usually considered non-hazardous.
- Much smaller space requirements than a chemical treatment system. A system that treats 10–20 gpm can easily fit in approximately a 4×10 footprint.
- Lower overall operation cost.

(vii) **Adsorption**

Adsorption is a mass transfer process which includes the accumulation of substances at the interface of two phases such as liquid-solid interface. In a solid–liquid framework adsorption brings about the expulsion of solutes from arrangement and their aggregation at strong surface. The solute staying in the arrangement achieves a dynamic equilibrium with that adsorbed on the solid phase (Thines et al. 2017). In the event that the collaboration between the strong surface and the adsorbed particles has a physical nature, the procedure is called physisorption. For this situation, the fascination interactions are van der Waals forces and, as they are frail the procedure results are reversible. Moreover, it happens lower or near the basic temperature of the adsorbed substance. Then again, if the fascination compels between adsorbed molecules and the strong surface are because of chemical bonding, the adsorption process is called chemisorptions (Krishnamurthy and Agarwal 2013).

For any adsorption procedure, an adsorbent having large surface territory, pore volume, and legitimate functionalities is the way to achievement. At present, a wide range of permeable materials have been created, for example, activated carbon, pillared muds, zeolites, mesoporous oxides, polymers and metal-natural structures, indicating shifting degree of adequacy in expelling the poisonous contaminations from air, water and soil. Among them, carbon-based adsorbents counting activated carbon, carbon nanotubes, for the most part show high adsorption limit and warm solidness. Of the different nanomaterials based adsorbents, carbon based materials have been tested as predominant adsorbents for the expulsion of inorganic and natural contaminants.

Activated carbon is a standout amongst the best media for evacuating an extensive variety of contaminants from industrial and metropolitan waste waters, landfill leachate and defiled groundwater. The activated carbon is delivered from different raw materials and, subsequently, it presents diverse properties.

Activated carbons (AC) (both granular activated carbon (GAC) and powdered activated carbons (PAC)) are normal adsorbents utilized for the evacuation of undesirable smell, shading, taste, and other organic and inorganic pollutions from residential and modern waste water attributable to their vast surface territory, smaller scale permeable structure nonpolar character and because of its financial viability. The real constituent of activated carbon is the carbon that records up to 95% of the mass weight. The dynamic carbons contain other hetero atoms, for example, hydrogen, nitrogen, sulphur, and oxygen. These are imitative from the source crude material or get to be related with the carbon amid activation and other planning techniques.

(a) Low Cost Adsorbent

Although, activated carbon is undoubtedly considered as all inclusive adsorbent for the evacuation of various sorts of pollutants from water, its broad utilization is here and they are confined because of the high expenses (Ahmaruzzaman 2008; Bhatnagar and Jain 2005). Endeavours have been made to grow minimal effort elective adsorbents which might be ordered in two ways it is possible that (i) on premise of their accessibility, i.e., (a) characteristic materials (wood, peat, coal, lignite and so forth.), (b) mechanical/rural/household squanders or by-items (slag, sludge, fly powder, bagasse fly ash, red mud etc.), and (c) combined items; or (ii) contingent upon their inclination, i.e., (a) inorganic and (b) organic material (Bhatnagar and Sillanpaa 2010; Gupta et al. 2009; Ahmaruzzaman 2008; Wan Ngah and Hanafiah 2008).

(b) Factors Affecting Adsorption Process

The components influencing the adsorption procedure are: (i) initial concentration of adsorbate (ii) solution pH (iii) surface area (iv) adsorbent dosage (v) temperature and (vi) interfering substances. Since adsorption is a surface phenomenon, the degree of adsorption is relative to the particular surface territory which is characterized as that bit of the aggregate surface territory that is accessible for adsorption (Naeem et al. 2007).

The physicochemical nature of the adsorbent definitely influences both rate and limit of adsorption. The solvency of the solute significantly impacts the adsorption equilibrium. As a rule, a converse relationship can be normal between the degree of adsorption of a solute and its dissolvability in the dissolvable where the adsorption happens. Atomic size is additionally important as it identifies with the rate of take-up of organic solutes through the permeable of the adsorbent material if the rate is controlled by intraparticle transport (Yu et al. 2009; Zhang and Huang 2007). For this situation the reaction will generally proceed more rapidly with decrease of adsorbate molecule.

The pH adjustment influences the degree of adsorption. Because of, the circulation of surface charge of the adsorbent can change (as a result of the creation of raw materials and the method of activation) accordingly changing the degree of adsorption as per the adsorbate functional groups (Putra et al. 2009; Gao and Pedersen 2005). Another critical parameter is the temperature. Adsorption responses are typically exothermic; thus the extent of adsorption generally increases with diminishing temperature (Onal et al. 2007; Bekci et al. 2006).

(viii) **Advanced Oxidation Process**

The instrument of AOP is the creation of OH radicals which are equipped for wrecking parts that are difficult to be oxidized. Era of OH radicals are the most part quickened by the blends of H_2O_2, UV, O_3, TiO_2, Fe^{2+}, electron bar light and ultra sound. AOPs are arranged under chemical, photochemical, catalytic, photo catalytic, mechanical and electrical processes. For the most part these procedures are found to decrease 70–80% of COD when contrasted with 30–45% lessening in the organic treatment.

(ix) **Wet Air Oxidation**

Wet air oxidation is a strategy for treating colour wastewater. In this propelled oxidation process, pure oxygen changes constituents (toxins) by oxidation, into carbon dioxide and water under high temperatures. The distinctions in the oxidation are decided in light of the medium at which oxidation is finished. Most normal mediums incorporate air or oxygen, which utilizes immaculate oxygen as a medium where in this case, air has been utilized for the development of oxidation items (Lei et al. 2000).

(x) **Other Advanced Oxidation Process**

Hydrogen peroxide/pyridine/copper(II) [H_2O_2/pyridine/Cu(II)] is a progressed oxidation technique which utilizes copper particles that act as a chelating specialist (pyridine) with the end goal of shaping anions and hydroxyl radicals. The copper particles are decreased to Cu(II) through this application. This treatment technique is fit for being utilized with the end goal of decolorizing the wastewater. The treatment strategy science starts with the utilization of copper(II) which through the chelating of pyridine changes in builds up a pyridine copper (II) complex (Gonder and Barlas 2005).

5 Sewerage Framework

With a specific end goal to effectively finish these procedures, there are some efficient systems which gather and transport the wastewater from the generation site to the treatment site. As specified before, without having this sort of sewer frameworks, some irresistible illnesses can spread out in public thus the sewerage framework is comprised of systems of various estimated channels (a portion of the funnels are big to the point that people can stroll through them), upkeep openings, pump stations, and trunk sewers. A sewerage framework fundamentally gets the squanders which have been discharged from mechanical or residential sources, and afterward the treatment strategies occur in the framework and it at last discharges the left-over poisons into the earth (Halus 1999). Moreover, the sewerage framework can move the water in basic circumstance like substantial rainfalls; thus, the surge does the negligible harm to the nationals and household zones. There are some imperative variables which must be considered before planning a sewerage treatment arrange, these comprise of.

5.1 Environmental Aspects

Environmental factors are ideas identified with ecological circumstances, for example, the nature of the groundwater and surface water, public health considerations, odour and insects disturbance which affect public health and land values.

5.2 Engineering Aspects

Engineering factors ought to be considered and seen by experienced architects to keep any basic marvels. The architects who take a shot at these activities ought to consider sewer infiltration, groundwater profundity, dangers of isolating, reusing of sewage inside the houses, bearing limit of the dirt, geology of the site and water powered computations, particularly for the beach front release.

5.3 Price Deliberation

Cost consideration is identified with a money related analyser who ought to foresee the beneficial perspectives which is the most important aspect in the sewage/squander water treatment framework. The analyser ought to focus on the costs incorporated into the venture which comprise of support, repair, power, fills, staff, compound and ought to likewise consider the general costs for types of gear, development and land cost.

6 Discarding/Recycling

Discarding or recycle are the last systems in sewage/wastewater treatment which ought to be outlined in a way which satisfies financial and scientific objectives and systems. There are two principle gatherings of transfer frameworks which vary in their utilization as far as where they are arranged and their capacity; in the on-site and Off-site (Public) transfer/reuse frameworks. On-site framework is being utilized as a part of a zone which has negligible pollution production and furthermore where there are just two or three houses in a wide geometric range. For this situation this transfer framework is considerably more beneficial regarding financial matters, in contrast with utilizing a framework for gathering, exchanging and treating the wastewater (Davis and Cornwell 2008).

In an open framework, squanders are being gathered in a region and after that conveyed to different areas through transporting frameworks for final disposal. Also open transfer frameworks can be exceptionally valuable in basic circumstances like flooding or tempests by containing the water in the off-site offices or completing the tremendous extent of the water.

7 Conclusion

The estimation of water assets is generally perceived and the personal satisfaction relies upon the capacity to oversee accessible water in the more prominent enthusiasm of the general population. The procedures of generation of materials particularly wet medicines and completing procedures of materials (completing the process of, colouring, printing, and so forth.) are enormous buyers of water with high calibre. As an after-effect of these different procedures, significant measures of dirtied water are discharged. The reuse of industry emanating is accepting more noteworthy significance in later times fundamentally because of shortage of water assets and progressively stringent administrative prerequisites for the transfer of the emanating. With reasonable treatment, profluent can be made fit for reuse or reuse in the generation procedure for which innovative alternatives are accessible. The profits by virtue of water and compound recuperation can offset working expense of gushing treatment reusing framework. The reusing and reuse of the treated profluent straightforwardly moderate common assets and a stage towards manageable improvement. Joints endeavours are required by water technologists and industry specialists to diminish water utilization in the business. While the client ventures ought to attempt advance water utilization, water technologists ought to embrace an incorporated way to deal with treat and reuse water in the industry. End-of-pipe innovations are utilized for wastewater treatment and incorporate consecutive use of an arrangement of techniques: coagulation/flocculation, flotation, adsorption, evaporation, oxidation, ignition, utilization of films, and so on that has been adjusted to the specific circumstance of an industrial plant. Therefore of the outrageous assortment of textile procedures and items, it is difficult to build up a sensible idea for a viable treatment of wastewater without a detailed investigation of the real circumstance in the industrial plant. To improve treatment and reuse potential outcomes, industrial waste streams ought to be on a fundamental level considered independently. At the point when the qualities of the different streams are known, it can be chosen which streams might be consolidated to make strides treatability and increment reuse alternatives. It is vital to investigate all parts of decreasing discharges and waste items from the industrial business since it will come about not just in enhanced ecological execution, in any case, additionally in generous funds on person organizations.

References

Ahmaruzzaman MD (2008) Adsorption of phenolic compounds on low-cost adsorbents: a review. Adv Colloid Interf Sci 143:48–67

Al-Bastaki N, Banat F (2004) Combining ultrafiltration and adsorption on bentonite in a onestep process for the treatment of colored waters. Resour Conserv Recy 41:103–113

Bekci Z, Seki Y, Yurdakoc MK (2006) Equilibrium studies for trimethoprim adsorption on montmorillonite KSF. J Hazard Mater B133:233–242

Bhatnagar A, Jain AK (2005) A comparative adsorption study with different industrial wastes as adsorbents for the removal of cationic dyes from water. J Colloid Interf Sci 281:49–55

Bhatnagar A, Sillanpaa M (2010) Utilization of agro-industrial and municipal waste materials as potential adsorbents for water treatment—a review. Chem Eng J 157:277–296

Bhatnagar A, Kumar E, Sillanpaa M (2011) Fluoride removal from water byadsorption—a review. Chem Eng J 811–840

Bidhendi GRN, Torabian A, Ehsani H, Razmkhah N, Abbasi M (2007) Evaluation of industrial dyeing wastewater treatment with coagulants. Int J Environ Res. 1:242–247

Bratby J (2006) Coagulation and flocculation in water and wastewater treatment. IWA Publishing, London

Choi HJ, Lee SM (2015) Heavy metal removal from acid mine drainage by calcined eggshell and microalgae hybrid system. Environ Sci Pollut Res 22:13404–13411

Davis ML, Cornwell DA (2008) Introduction to environmental engineering. McGraw-Hill Companies, New York, p 456

Donald R, Rowe IMAM (1995) Handbook of wastewater reclamation and reuse. CRC Press

Gao J, Pedersen JA (2005) Adsorption of sulfonamide antimicrobial agents to clay minerals. Environ Sci Technol 39:9509–9516

Geise GM, Lee HS, Miller DJ, Freeman BD, McGrath JE, Paul DR (2010) Water purification by membranes: the role of polymer science. J Polym Sci, Part B: Polym Phys 1685–1718

George JS, Ramos A, Shipley HJ (2015) Tanning facility wastewater treatment: analysis of physical-chemical and reverse osmosis methods. J Environ Chem Eng 3:969–976

Gonder B, Barlas H (2005) Treatment of coloured wastewater with the combination of Fenton process and ion exchange. Fresenius Environ Bull 14:393–399

Gupta VK, Carrott PJM, Ribeiro Carrott MML, Suhas TL (2009) Low-cost adsorbents: growing approach to wastewater treatment—a review. Crit Rev Env Sci Technol 39:783–842

Halus WJ (1999) Sewage water purification/reuse/redistribution, food control, and power generating system. US Patent 6,000,880

Harremoes CJAH (2002) Wastewater treatment biological and chemical processes

Ihsanullah, Abbas A, Al-Amer AM, Laoui T, Al-Marri M, Nasser MS, Khraisheh M, Atieh MA (2016) Heavy metal removal from aqueous solution by advanced carbon nanotubes: critical review of adsorption applications. Sep Purif Tech 157:141–161

Jiang L, Luo S, Fan X et al (2011) Biomass and lipid production of marine microalgae using municipal wastewater and high concentration of CO_2. Appl Energy 88:3336–3341

Kawamura S (2000) Integrated design and operation of water treatment facilities. Wiley

Krishnamurthy G, Agarwal S (2013) Optimization of reaction conditions for high yield synthesis of carbon nanotube bundles by low-temperature solvothermal process and study of their H_2 storage capacity. Bull Korean Chem Soc 34(10):3046–3054

Kumar P, Pournara A, Kim KH, Bansal V, Rapti S, Manos MJ (2017) Metal-organic frameworks: challenges and opportunities for ion-exchange/sorption applications. Prog Mat Sci 86:25–74

Lei L, Hu X, Chen G et al (2000) Wet air oxidation of desizing wastewater from the textile industry. Ind Eng Chem Res 39:2896–2901

Lettinga G, Pol LH (2008) New technologies for anaerobic wastewater treatment, 2nd edn. Wiley

Mehta D, Mazumdar S, Singh SK (2015) Magnetic adsorbents for the treatment of water/ wastewater—a review. J Water Process Eng 7:244–265

Metcalf E, Eddy H (1987) Wastewater engineering: treatment, disposal, reuse. Tata McGraw-Hill Publishing Company Ltd

Naeem A, Westerhoff P, Mustafa S (2007) Vanadium removal by metal (hydr)oxide adsorbents. Water Res 41:1596–1602

Onal Y, Akmil-Basar C, Sarici-Ozdemir C (2007) Elucidation of the naproxen sodium adsorption onto activated carbon prepared from waste apricot: kinetic, equilibrium and thermodynamic characterization. J Hazard Mater 148:727–734

Oreopoulou V, Russ W (2006) Utilization of by products and treatment of waste in the food industry. Springer Science and Business Media

Paul SA, Chavan SK, Khambe SD (2012) Studies on characterization of textile industrial waste water in Solapur city. Int J Chem Sci 10:635–642

Putra EK, Pranowo R, Sunarso J, Indraswati N, Ismadji S (2009) Performance of activated carbon and bentonite for adsorption of amoxicillin from wastewater: mechanisms, isotherms and kinetics. Water Res 43:2419–2430

Qasim SR (1998) Wastewater treatment plants: planning, design, and operation. CRC Press

Qu X, Alvarez PJJ, Li Q (2013) Applications of nanotechnology in water and wastewater treatment. Water Res 47:3931–3946

Raeesossadati MJ, Ahmadzadeh H, Mchenry MP et al (2014) CO_2 bioremediation by microalgae in photobioreactors: impacts of biomass and CO_2 concentrations, light, and temperature. Algal Res 6:78–85

Rao MM, Ramesh A, Rao GPC, Seshaiah K (2006) Removal of copper and cadmium from the aqueous solutions by activated carbon derived from *Ceiba pentandra* hulls. J Hazard Mater 129:123–129

Rosen TWM, Lofqvist A (1998) Development of a new process for treatment of a pharmaceutical wastewater

Sahu AK, Siljudalen J, Trydal T et al (2013) Utilisation of wastewater nutrients for microalgae growth for anaerobic co-digestion. J Environ Manage 122:113–120

Somasiri W, Li X, Ruan W et al (2008) Evaluation of the efficacy of upflow anaerobic sludge blanket reactor in removal of colour and reduction of COD in real textile wastewater. Bioresour Technol 99:3692–3699

Thines RK, Mubarak NM, Nizamuddin S et al (2017) Application potential of carbon nanomaterials in water and wastewater treatment: a review. J Taiwan Inst Chem Eng 000:1–18

Tilley DF (2011) Aerobic wastewater treatment processes: history and development. IWA Publishing

Verma AK, Dash RR, Bhunia P (2012) A review on chemical coagulation/flocculation technologies for removal of colour from textile wastewaters. J Environ Manage 93:154–168

Wan Ngah WS, Hanafiah MAKM (2008) Removal of heavy metal ions from wastewater by chemically modified plant wastes as adsorbents: a review. Bioresour Technol 99:3935–3948

Wijetunga S, Li X, Jian C (2010) Effect of organic load on decolourization of textile wastewater containing acid dyes in upflow anaerobic sludge blanket reactor. J Hazard Mater 177:792–798

Wong PW, Teng TT, Norulaini NARN (2007) Efficiency of the coagulation-flocculation method for the treatment of dye mixtures containing disperse and reactive dye RID F-6428-2010. Water Qual Res J Can 42:54–62

Wu TY, Mohammad AW, Lim SL et al (2013) Recent advances in the reuse of wastewaters for promoting sustainable development. In: Sharma SK, Sanghi R (eds) Wastewater reuse and management. Springer, Netherlands, pp 47–103

Yu Z, Peldszus S, Huck PM (2009) Adsorption of selected pharmaceuticals and an endocrine disrupting compound by granular activated carbon. 2. Model prediction. Environ Sci Technol 43:1474–1479

Zhang H, Huang CH (2007) Adsorption and oxidation of fluoroquinolone antibacterial agents and structurally related amines with goethite. Chemosphere 66:1502–1512

Review of Utilization Plant-Based Coagulants as Alternatives to Textile Wastewater Treatment

Thabata Karoliny Formicoli Souza Freitas, Cibele Andrade Almeida, Daniele Domingos Manholer, Henrique Cesar Lopes Geraldino, Maísa Tatiane Ferreira de Souza and Juliana Carla Garcia

Abstract With the increased demand for textile products, the textile industry and its wastewaters have been increasing proportionally, making it one of the main sources of severe pollution problems worldwide. Textile wastewater treatment is one the most difficult environmental problems because it contains high color, biochemical oxygen demand (BOD), chemical oxygen demand (COD), pH, temperature, turbidity and toxic chemicals. The direct discharge of this wastewater without previous or proper treatment into the water bodies, like lakes, rivers, etc. pollutes the water affecting directly and indirectly the water. Coagulation/flocculation is one of the most widely used for wastewater treatment, as it is efficient and simple to operate. This process is used for the removal of suspended and dissolved solids, colloids and organic matter present in industrial wastewater. Natural coagulants have been attracting wide interest of researchers because they have the advantages of biodegradability, safe for human health, environmental friendly, generally toxic free and produce no secondary pollution. These coagulants are extracted from natural and renewable sources, such as microorganisms, animals or plants. Not only this, the sludge volume generated by the natural coagulants is smaller than chemical coagulants; it can further be treated biologically or can be disposed safely as soil conditioners because of their non-toxicity. The raw plant extracts are often available

T.K.F.S. Freitas · C.A. Almeida · D.D. Manholer · H.C.L. Geraldino · M.T.F. de Souza · J.C. Garcia (✉)
GPDMA, Department of Chemistry, State University of Maringá, Maringá, Paraná, Brazil
e-mail: jucgmoraes@uem.br

T.K.F.S. Freitas
e-mail: thabata.karoliny@gmail.com

C.A. Almeida
e-mail: cibelealmeida2@hotmail.com

D.D. Manholer
e-mail: dmanholer@hotmail.com

H.C.L. Geraldino
e-mail: henriqueclg@hotmail.com

M.T.F. de Souza
e-mail: maisataty@hotmail.com

© Springer Nature Singapore Pte Ltd. 2018
S.S. Muthu (ed.), *Detox Fashion*, Textile Science and Clothing Technology,
DOI 10.1007/978-981-10-4780-0_2

locally and hence, a low-cost alternative to chemical coagulants. In recent years, numerous studies on natural coagulants are growing and there is an urgent need to establish the use of natural low-cost coagulants for textile wastewater treatment. In this chapter, we show the characteristics of dyes and textile wastewater, emphasizing adverse impacts on environmental and human health and we mentioned some technologies for the textile wastewater treatment, highlighting the CF, since it is efficient, is easy to operate and is commonly used at the industries. We also have been discussed the physical-chemical concept of CF as well the major mechanisms involved at process. The usage of plant-based natural coagulants as alternative to chemical coagulants in the textile wastewater treatment is the goal of this chapter.

Keywords Natural coagulant · Plant coagulant · Coagulation/flocculation · Textile treatment · Textile effluent

List of Abbreviations

ANOVA   Analyses of variance
BOD   Biochemical oxygen demand
CF   Coagulation/flocculation process
COD   Chemical oxygen demand
OD   Oxygen dissolved
PAC   Polyaluminium chloride
POPs   Persistent organic pollutants
TCD   Textile contact dermatitis
TDS   Total dissolved solids
TDSO   Tumorigenic dose rate 50
TOC   Total organic carbon
TSS   Total suspense solids
USEPA   Environmental Protection Agency

1   Textile Wastewater

Environmental pollution is currently one of the most important issues faced by humanity. Pollution has increased exponentially in the past few years and has reached alarming levels in terms of its effects on living creatures (Prasad and Rao 2016). Increase in demand for rapid urbanization and changing consumption patterns, together with unrestrained population growth, rapid socioeconomic development and changes in agriculture, medicine, energy sources, and chemical industries have all resulted in the generation of environmental pollutants. These pollutants are released into air, water, and soil and have detrimental effects on the health of humans, plants, animals, and microbes (Ali 2010; Colpini et al. 2014; Souza et al. 2016; Uday et al. 2016).

The textile industry is one of the most important manufacturing sectors and produces large volumes of highly toxic wastewater due to the high quantities of water, dyes and chemicals used in its processes (Jegatheesan et al. 2016; Gümüş and Akbal 2011). The World Bank estimates that 17–20% of industrial water pollution is contributed by the textile industry (Kant 2012; Jegatheesan et al. 2016).

Scouring, dyeing, printing, finishing and washing processes contribute the highest volumes of wastewater (Dasgupta et al. 2015). The traditional textile finishing industry consumes about 100 L of water per 1 kg of textile materials (Ali 2010). The dyeing industry consumes about 30–50 L of water per 1 kg of cloth depending on the type of dye used. The overall water consumption of yarn dyeing is about 60 L per 1 kg of yarn (Kant 2012). In general, textile industries typically generate 200–350 L of wastewater per 1 kg of finished product (Gozálvez-Zafrilla et al. 2008) resulting in an average pollution of 100 kg chemical oxygen demand (COD) per ton of fabric (Jekel 1997). For a textile unit processing 400,000 lb of cotton per week, more than 50,000 lb of salts are released. The usual salt concentration in wastewater is 2000–3000 ppm (Khandegar and Saroha 2013). The salts in the effluent can lead to the soil infertility and aquatic life damage (Khandegar and Saroha 2013). Not to mention that about 2 to 50% of the total dyes that have not been fixed to the fiber are lost during dyeing and washing process, leading to severe contamination of surface and ground water (O'Neill et al. 1999).

The Environmental Protection Agency (USEPA) has classified textile wastes into four principal groups, namely dispersible, hard-to-treat, high-volume, and hazardous and toxic wastes (Foo and Hameed 2010). The composition and volume of wastewater from textile industries exhibit wide heterogeneity, depending on many different factors, including the type of fibers, chemicals and dyes, machines, techniques, the season of the year, the characteristic quality imparted to processed fabric, the nature of the special finishing if any, the specificity of the process and the principles on which the water use has been modeled (Santos et al. 2007; Dasgupta et al. 2015; Soares et al. 2017).

In general, textile wastewater is quite a complex mixture and is highly variable, comprising many polluting substances. It is a mixture of different types of dyes and auxiliary products such as surfactants, fixing agents, oxidizing agents, recalcitrant chlorinated compounds, salts, heavy metals, dispersing agents and smoothing agents (compounds that may be both persistent and toxic) (Hassani et al. 2008; Zahrim et al. 2011). Wastewaters are highly colored due to the presence of dyes that have not been fixed to the fiber during the dyeing process (Kunz et al. 2002). They are usually also characterized by high levels of chemical oxygen demand (COD), biochemical oxygen demand (BOD), pH, salinity, temperature, turbidity, toxic chemical compounds, total dissolved solids (TDS) and total suspended solids (TSS) (Valh et al. 2011; Verma et al. 2012). Furthermore, many textile dyes or by-products are themselves toxic, carcinogenic, mutagenic and/or teratogenic (Mathur et al. 2005; Khandegar and Saroha 2013; Santos et al. 2013). Table 1 lists the characteristics of textile wastewaters from different stages of textile industry processes, and wastewater discharge standards prescribed by the statutory authorities.

Table 1 Physical-chemical characteristics of textile wastewater from textile processing (Santos et al. 2007; Valh et al. 2011; Khandegar and Saroha 2013; Arslan et al. 2016)

Parameters	General industry	Scouring	Bleaching	Mercerising	Dyeing	Composite	Desizing	Discharge limit into public sewage (Indian legislation)	Discharge limit into water (Brazilian legislation)	Discharge limit into water and municipal wastewater treatment plan (China legislation)	
										Direct	Indirect
pH	1.9–13	10–12	8.5–11	8–10	9–11	8–10	–	5.6–9.0	5.0–9.0	6.0–9.0	6.0–9.0
TDS (mg L^{-1})	2900–3100	12,000–30,000	2500–11,000	2000–2600	1500–4000	5000–10,000	16,000–32,000	2100	–	–	–
TSS (mg L^{-1})	15–64,000	1000–2000	200–400	100–400	50–350	100–700		100	–	60	100
BOD$_5$ (mg L^{-1})	50–40,000	2500–3500	100–500	50–120	100–400	50–550	1700–5200	30	50	25	50
COD (mg L^{-1})	150–12,000	10,000–20,000	1200–1600	250–400	400–1400	250–8000	4600–5900	250	200	100	200
Color	50–2500 (mgPtCo L^{-1})	–	–	Highly colored	Strongly colored	Strongly colored	–	Colorless	75 (mgPtCo L^{-1})	70 (dilution ratio)	80 (dilution ratio)

COD: chemical oxygen demand, BOD$_5$: biochemical oxygen demand incubated to 5 days, mgPtCo L^{-1}: milligrams Platinum-Cobalt per liter, TDS: total dissolved solids, TSS: total suspended solids

*Brazilian legislation, Conama Resolution n° 430/2011 and CEMA Resolution n° 70/2009; China legislation, GB 4287-2012; Bureau of Indian Standards, BIS-3306-1955

2 Textile Dyes

Colorants (dyes and pigments) are important industrial chemicals. Pigments are colorants which are insoluble in water and do no interact with the substrates, whereas dyes are mostly soluble in water and diffuse into the material and are fixed to colorize the material (Chequer et al. 2011; Uday et al. 2016). Textile dyes are organic compounds that give color to fibers (substrate). Although in chemistry, "dye" denominates only substances of aromatic character that are capable of irreversibly coloring a textile support, the term "dye" can refer to any substance that gives color, including both dyes and paints (Moraes et al. 2000; Dellamatrice et al. 2008).

There are different types of classifications for dyes. They can be classified in terms of their chemical structure, color and application methods. However, due to the complexities of color nomenclature based on the chemical structure system, classification based on the application is often favorable (Guaratini and Zanoni 2000; Yagub et al. 2014). Fibers can take up dyes as a result of van der Waals forces, hydrogen bonds and hydrophobic interactions. The uptake of the dye into fibers depends on the nature of the dye and its chemical constituents. The strongest dye-fiber attachments are as a result of covalent bonds with additional electrostatic interactions where the dye ions and the fibers have opposite charges (Santos et al. 2007). Thus, knowing the nature of the fiber is important because the fiber structure determines the kind of dye that should be used and application method (Moura 2003). Figure 1 shows the main types of dyes used for dyeing different kinds of fiber and Table 2 represents the characteristics of different dyes that are used widely in the textile industry. It can be seen in Table 2 that reactive dyes are widely used to color cotton, which contributes half of the worldwide textile-fiber market (Verma et al. 2012).

Dyes are also classified based on their chemical structure and they are constituted by two main groups, chromophores and auxochromes (based on their functional groups). Figure 2 illustrates the chemical structure of dye molecules. Chromophore groups are characterized by the presence of a delocalized electron system with conjugated double bonds (π-π* and n-π* transitions) in the molecular

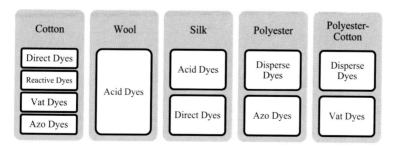

Fig. 1 Dyes for different fibers. Reprinted from Holkar et al. (2016) with permission of Copyright © 2016 Elsevier Ltd.

Table 2 Classification of dyes mainly used in textile industry and its method of application (O'Neill et al. 1999; Guaratini and Zanoni 2000; Hunger 2007; Verma et al. 2012; Yagub et al. 2014)

Class	Characteristics	Substrate (Fiber)	Dye-fiber interaction	Application method	Chemical types	Degree of fixation (%)	Loss to effluent (%)
Acid	Anionic, water soluble	Wool, nylon, silk	Electrostatic, Hydrogen bonding	Generally from neutral to acidic dyebath	Anthraquinone, azine, xanthene, azo (including, nitroso, premetallised), nitro, triphenylmethane, xanthene	89–95	5–20
Basic	Cationic, water soluble	polyacrylonitrile, treated nylon, and polyester	Electrostatic attraction	Applied from acidic dyebaths	Hemicyanine, azo, cyanine, azine, diazahemicyanine, diphenylmethane, xanthene, acridine, triarylmethane, anthraquinone, oxazine	95–100	0–5
Direct	Anionic, water soluble	Nylon, rayon, leather and cotton	Intermolecular forces	Applied from neutral or a little alkaline bath containing additional electrolyte	Phthalocyanine, azo, oxazine, stilbene	70–95	5–30
Disperse	Very low water solubility	Polyamide, acrylic polyester and acetate	Hydrophobic-solid state mechanism	Fine aqueous dispersions often applied by high temperature/pressure or lower temperature carrier methods; dye may be padded on cloth and thermo fixed	Benzodifuranone, azo, anthraquinone, nitro, styryl	90–100	0–10

(continued)

Table 2 (continued)

Class	Characteristics	Substrate (Fiber)	Dye-fiber interaction	Application method	Chemical types	Degree of fixation (%)	Loss to effluent (%)
Reactive	Anionic, water soluble	Wool, cotton, silk and nylon	Covalent bonding	Reactive site on dye reacts with functional group on fiber to bind dye covalently under influence of heat and pH (alkaline)	Anthraquinone, formazan, phthalocyanine, azo, oxazine and basic	50–90	10–50
Sulphur	Colloidal, insoluble	Rayon and cotton	Covalent bonding	Aromatic substrate vatted with sodium sulphide and re-oxidised to insoluble Sulphur—containing products on fiber	Indeterminate structures	60–90	10–40
Vat	Colloidal, insoluble	Wool and cotton	Impregnation and oxidation	Water-insoluble dyes solubilized by dropping in sodium hydrogen sulphite, then exhausted on re-oxidised and fiber	Indigoids and anthraquinone	80–95	5–20

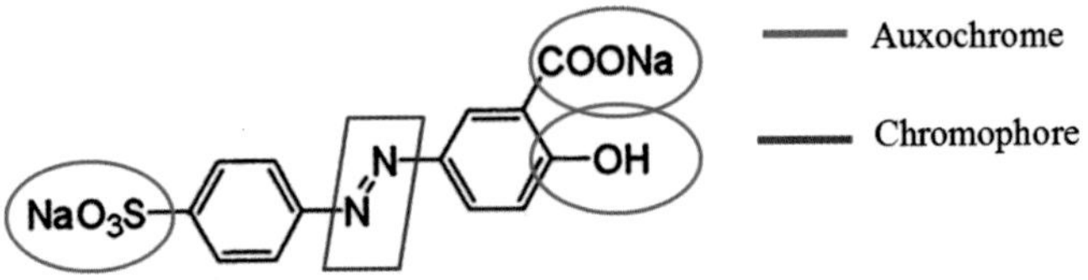

Mordant Yellow 10 (Azo dye)

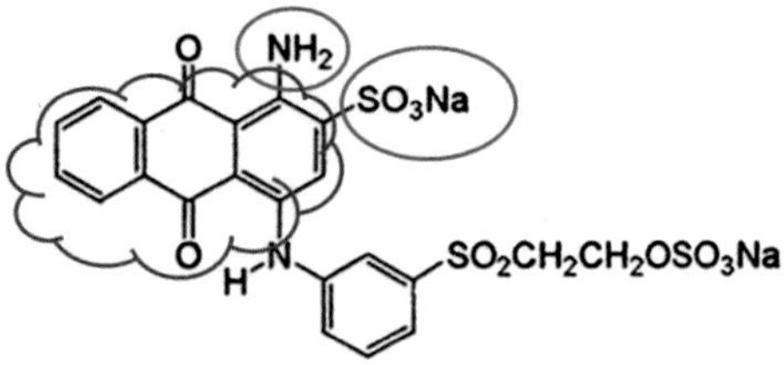

Remazol Brilliant Blue (Anthraquinone dye)

Fig. 2 Examples of dye-auxochromes and–chromophores for azo and anthraquinone dyes. Reprinted with permission of Santos et al. (2007). Copyright © 2016 Elsevier Ltd.

structure such as –C=C–, –C≡N–, –C=O–, –N=N– that are responsible for the dye colors. Functional groups are characterized by the presence of an electron withdrawing or donating substituent (auxochromes cannot undergo π-π^* transitions but can undergo transition of n electrons), which enhances the affinity of the dye toward the fibers, consequently enhancing the color intensity. The most common auxochromes are the $-NH_2$, $-NR_2$, –NHR, –COOH, $-SO_3H$, –OH and $-OCH_3$ (Kiernan 2001; Kunz et al. 2002; Colpini et al. 2008; Uday et al. 2016). It is worth mentioning that sulfonate groups confer very high aqueous solubility to dyes (Santos et al. 2007). Table 3 shows the classification of dyes based on the presence of chromophore groups.

Azo dyes constitute one of the most important groups of synthetic dyes used by textile industries (Holkar et al. 2016; Santos et al. 2016). Among commercial synthetic dyes, azo dyes constitute a group with a broad range of colors and structures and represent up to 70% of the all dyestuffs used worldwide. They are the most common synthetic dyes released into the environment (Saratale et al. 2011; Sen et al. 2016). Azo dyes absorb light in the visible spectrum due to their chemical structure, which is characterized by one or more azo group (–N=N–) bonds to aromatic rings (Fig. 3) (Castanho et al. 2006). The azo group is substituted with benzene or naphthalene groups, which can contain many different substituents, such as chloro (–Cl), methyl ($-CH_3$), nitro ($-NO_2$), amino ($-NH_2$), hydroxyl (–OH), or carboxyl (–COOH), giving different types of azo dyes (Zollinger 1991). The success of azo dyes is due to their simple synthetic procedures, their great structural diversity, their high molar extinction coefficient, and their medium-to-high fastness properties with respect to both light and wetness (Bafana et al. 2011).

Table 3 Classification of dyes based on the chromophore present (Ali 2010)

Class dyes	Chromophore	Example
Nitro		 C.I. Acid Yellow 24
Nitroso		 Fast Green O
Azo		 Methyl Orange
Triphenylmethane		 C.I. Basic Violet 3
Phthalein		 Phenolphthalein

(continued)

Table 3 (continued)

Class dyes	Chromophore	Example
Indigo		C.I. Acid Blue 71
Anthraquinone		C.I. Reactive Blue 19

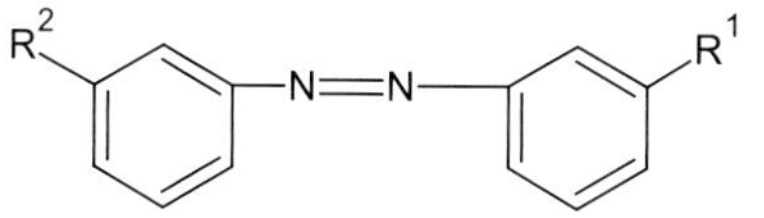

Fig. 3 Basic structure of azo dye. Reprinted with permission of Khan et al. (2013). Copyright © (2012) Springer Science Business Media B.V.

Improper discharge of textile wastewater containing azo dyes and their metabolites into water bodies can cause serious environmental problems, such as visual pollution due to their high coloration, inhibition of photosynthetic processes as result of reduced sunlight penetration, deficiency of dissolved oxygen concentration, changes the life cycle of aquatic organisms and toxicity to aquatic flora and fauna, even at low dye concentrations (Foo and Hameed 2010; Saratale et al. 2011; Simionato et al. 2014; Góes et al. 2016).

3 Effects of Textile on Environmental and Human Health

Depending on the concentration and exposure time, dyes can have acute and/or chronic effects on exposed organisms (Valh et al. 2011). Today, two classes of dyes, azo and disperse, are recognized worldwide as having the potential to cause allergic contact dermatitis and possibly cancer. They can be cleaved by skin bacteria, or by dermal or systemic metabolism into aromatic amines which might have undesired toxicological properties, particularly due to their carcinogenic or allergenic potential (Platzek 2010; Brüschweiler et al. 2014). After a detailed literature search of 43 high priority substances, Brüschweiler et al. (2014) organized two priorities lists. Priority list A (Table 4) contains 15 (potentially) carcinogenic and/or genotoxic substances; priority list B (Table 5) contains 11 substances that may (potentially) cause sensitization by skin contact.

Textile contact dermatitis (TCD) results from contact between the skin and substances used in clothing. Finishing products (finish resins) and dyes are the main causes of TCD. During contact there can be perspiration, which involves moisture transport between the skin and the dyed and finished clothing items. These sensitizing agents (they can be fibers, dyes, or finish resins) are transported by transpiration into the skin and can cause skin irritation, including eczematous dermatitis (in most cases), urticaria, purpura, erythema multiforme, and erythroderma (Brookstein 2009; Sánchez-Gilo et al. 2010).

Besides causing contact allergies and dermatitis, if toxic dyes are able to penetrate the skin and enter the body, they can also induce other kinds of harmful effects to the exposed cells. The possible deleterious effects include DNA damage, kidney and pancreatic problems, nervous system damage and even cancer (Leme et al. 2014).

Table 4 Priority list A: (Potentially) carcinogenic and/or genotoxic aromatic amines (Brüschweiler et al. 2014)

Cleavage product	CASRN	Chemical structure	Points of concern	Azo dye parent compound)
2,6-Xylidine 2,6-Dimethylaniline	87-62-7	(NH₂ structure)	IARC Carc. Cat. 2B (possible human carcinogen) EU Carc. Cat. 3; R40 (limited evidence of a carcinogenic effect) MAK Carc. Cat. 2 (DFG)	C.I. Acid Orange 134
2,4-Xylidine 2,4-Dimethylaniline	95-68-1	(NH₂ structure)	MAK Carc. Cat. 2 (DFG)	C.I. Acid Red 8 C.I. Acid Red 26 C.I. Acid Red 40 C.I. Acid Red 135 C.I. Acid Red 170 C.I. Acid Orange 17 C.I. Direct Red 126 C.I. Direct Red 168 C.I. Solvent Orange 7 C.I. Solvent Yellow 18

(continued)

Table 4 (continued)

Cleavage product	CASRN	Chemical structure	Points of concern	Azo dye parent compound)
3,4-Dichloroaniline	95-76-1		Limited genotoxicity testing produced equivocal results. Positive sister chromatid exchange test, evidence of induction of spindle damage, negative micronucleus test). No carcinogenicity study available	C.I. Disperse Yellow 241
2-Methoxy-4-nitroaniline 4-Nitro-o-anisidine	97-52-9		R40 (limited evidence of a carcinogenic effect). Positive in vitro genotoxicity tests (Ames-test, chromosomal aberration). It is on the NTP candidate list for carcinogenicity study	C.I. Disperse Red 41
2-Amino-4-nitrophenol	99-57-0		$TD50 = 839$ mg/kg bw/day (rat). Inadequate evidence in humans for the carcinogenicity. Limited evidence for the carcinogenicity in experimental animals. Not classifiable as to its carcinogenicity to humans (IARC group 3)	C.I. Acid Brown 349 C.I. Acid Yellow 119 C.I. Mordant Brown 33
2-Amino-4-nitroanisol 5-Nitro-o-anisidine	99-59-2		$TD_{50} = 53.9$ mg/kg bw/day (rat). $SFo = 0.049$ 1/(mg/kg bw/day) (US EPA). Positive in genotoxicity tests. Increased incidence of tumours in rats and mice. Not classifiable as to its carcinogenicity to humans	C.I. Pigment Red 23
4-Nitroaniline	100-01-6		$SFo = 0.02$ 1/(mg/kg bw/day) (US EPA). Significant increase in hepatic hemangiosarcoma in male mice. It recommended to classify it as suspected human carcinogen	C.I. Acid Black 1 C.I. Acid Black 210 C.I. Acid Brown 75 C.I. Acid Brown 106 C.I. Acid Brown 123

(continued)

Table 4 (continued)

Cleavage product	CASRN	Chemical structure	Points of concern	Azo dye parent compound)
				C.I. Acid Brown 348
				C.I. Acid Brown 349
				C.I. Disperse Orange 1
				C.I. Disperse Orange 1.1
				C.I. Disperse Orange 3
				C.I. Disperse Orange 25
				C.I. Disperse Orange 29
				C.I. Disperse Orange 31
				C.I. Disperse Orange 33
				C.I. Disperse Orange 49

(continued)

Table 4 (continued)

Cleavage product	CASRN	Chemical structure	Points of concern	Azo dye parent compound)
				C.I. Disperse Orange 73
				C.I. Disperse Orange 80
				C.I. Disperse Red 1
				C.I. Disperse Red 2
				C.I. Disperse Red 17
				C.I. Disperse Red 19
				C.I. Disperse Red 74
				C.I. Disperse Red 135
				C.I. Disperse Red 278
				C.I. Mordant Orange 1

(continued)

Table 4 (continued)

Cleavage product	CASRN	Chemical structure	Points of concern	Azo dye parent compound
2-Amino-5-nitrothiazol	121-66-4		TD50 = 44.6 mg/kg bw/day (rat) Limited evidence for the carcinogenicity in experimental animals	C.I. Disperse Blue 102 C.I. Disperse Blue 106 C.I. Disperse Blue 124
2-Amino-5-nitrophenol	121-88-0		$TD_{50} = 111$ mg/kg bw/day (rat) Inadequate evidence in humans for the carcinogenicity Limited evidence in experimental animals for the carcinogenicity Not classifiable as to its carcinogenicity to humans (IARC Group 3)	C.I. Disperse Red 16
4-Aminophenol	123-30-8		EU Muta. Cat. 3; R68 (possible risk of irreversible effects)	C.I. Acid Orange 33 C.I. Disperse Orange 13 C.I. Disperse Orange 29

(continued)

Table 4 (continued)

Cleavage product	CASRN	Chemical structure	Points of concern	Azo dye parent compound
2-Amino-6-methoxybenzothiazol	1747-60-0		Some evidence for genotoxicity	C.I. Disperse Red 379
4-Nitro-1,3-phenylenediamine	5131-58-8		Evidence for genotoxicity (some Ames-tests positive, chromosome aberration positive, sister chromatid exchange weakly positive)	C.I. Acid Brown 83
2-Amino-6-nitrobenzothiazol	6285-57-0		Positive Ames-tests	C.I. Disperse Red 177 C.I. Disperse Red 179
3-Amino-5-nitro-2,1-benzisothiazol	14346-19-1		Positive Ames-tests	C.I. Disperse Blue 148
2,5-Diaminotoluene p-Toluenediamine	25376-45-8		Increased tumours in liver, mammary gland and lung of rats EU Carc. Cat. 2; R45 (may cause cancer) EU Muta. Cat. 3; R68 (possible risk of irreversible effects) EU Repr. Cat. 3; R62 (possible risk of impaired fertility)	C.I. Solvent Red 25

TD50: tumorigenic dose rate 50. SFo: oral slope factor

Table 5 Priority list B: aromatic amines which may cause sensitization by skin contact (Brüschweiler et al. 2014)

Cleavage product	CASRN	Chemical structure	Points of concern	Azo dye parent compound
1,3-Phenylenediamine-4-sulfonic acid	88-63-1		R43 (may cause sensitization by skin contact)	C.I. Acid Brown 123
3,4-Dichloroaniline	95-76-1		R43 (may cause sensitization by skin contact)	C.I. Disperse Yellow 241
N,N-Dimethyl-1,4-Phenylenediamine N,N-Dimethyl-p-benzenediamine	99-98-9		Contact dermatites (HSDB)	C.I. Basic Blue 54 C.I. Disperse Red 4 C.I. Solvent Yellow 2 Janus Green B
4-Aminodiphenylamine	101-54-2		R43 (may cause sensitization by skin contact) Contact dermatites (HSDB)	C.I. Acid Yellow 36 C.I. Disperse Orange 1
p-Phenylenediamine	106-50-3		R43 (may cause sensitization by skin contact)	C.I. Acid Brown 4 C.I. Acid Brown 237 C.I. Acid Green 11 C.I. Acid Red 151 monolithium salt C.I. Acid Yellow 159 C.I. Direct Black 8 C.I. Direct Black 19 disodium salt C.I. Direct Red 81 C.I. Direct Red 253 C.I. Disperse Black 1 C.I. Disperse Black 2 C.I. Disperse Orange 3

(continued)

Table 5 (continued)

Cleavage product	CASRN	Chemical structure	Points of concern	Azo dye parent compound
Sulfanilic acid 4-Aminobenzene sulfonic Acid	121-57-3		R43 (may cause sensitization by skin contact)	C.I. Acid Black 210 C.I. Acid Brown 106 C.I. Acid Orange 7 C.I. Acid Orange 156 C.I. Acid Red 151 C.I. Direct Brown 44 C.I. Direct Red 80 C.I. Direct Red 81 C.I. Direct Red 253
4-Aminophenol	123-30-8		Known allergen (HSDB)	C.I. Acid Orange 33 C.I. Disperse Orange 13 C.I. Disperse Orange 29
4-Ethoxyaniline p-Phenetidine p-Aminophenetol	156-43-4		R43 (may cause sensitization by skin contact)	C.I. Acid Yellow 38 C.I. Acid Yellow 159 C.I. Direct Yellow 12
4-(N,N-Diethyl)-2-methyl-p-phenylenediamine monohydrochloride	2051-79-8		R43 (may cause sensitization by skin contact)	C.I. Acid Red 397 C.I. Disperse Red Dye
1,4-Diamino-2-methoxybenzene	5307-02-8		Contact dermatites is common (HSDB)	C.I. Acid Yellow 219 C.I. Disperse Orange 29
2,5-Diaminotoluene p-Toluenediamine	25376-45-8		R43 (may cause sensitization by skin contact) A potent skin sensitizer (HSDB)	C.I. Solvent Red 25

Textile dyes also affect human metabolism through the food chain, causing kidney and respiratory problems, and hypertension, among others. Studies carried out with workers in the textile industry indicated that these professionals are more prone to pancreatic cancer, bladder cancer and cancer of the digestive system because they are exposed to dyes on a daily basis (Peralta-Zamora et al. 2002; Aksu and Donmez 2005; Sanghi and Verma 2013).

Azo and nitro dyes are reduced in the intestinal environment, resulting in the formation of toxic amines in both cases (Chung et al. 1978; Weber and Wolfe 1987). Studies have linked exposure to dyes in textile workers and increased risk of bladder, digestive system and pancreatic cancer (Dolin 1992; Notani et al. 1993; Alguacil et al. 2000; Mastrangelo et al. 2002). Vineis and Pirastu (1997) reported that exposure of workers to aromatic amines increases the incidence of bladder cancer by 25%.

Bosetti et al. (2005) conducted a population-based study in Canada and found a positive association between bladder cancer and occupational exposure to paints in workers in the textile industry. De Roos et al. (2005) also observed an increase in the risk of colon cancer in textile workers exposed to dyes for a long period (20 years or more). Other studies also found an association between esophageal cancer and workers in the textile industry (Wernli et al. 2006).

Wastewaters from the textile industry are usually polluted with recalcitrant or hazardous organics, such as dyes, surfactants, metals, salts, and persistent organic pollutants (POPs) (Valh et al. 2011). These industries discharge large volumes of wastewaters into aquatic environments, which may affect aquatic organisms both directly and indirectly, through the trophic chain or by resuspension, which may make pollutants bioavailable and enable them to reach human beings (Garcia et al. 2012). Verma et al. (2012) summarized the harmful direct and indirect effects of textile wastewaters in the environment as shown in Fig. 4.

Although textile wastewaters are variable mixtures of many pollutants, their major characteristic is high coloration due to the presence of dyes. This high coloration coupled with high organic load induces perturbation to aquatic life and causes dramatic aesthetic pollution (Blanco et al. 2014; Gupta et al. 2015), which cannot be tolerated by local populations (Verma et al. 2012). The greatest environmental concern about dissolved dyes in water bodies is not only their visibility, but also because they absorb sunlight and can hinder light penetration in the water (rivers, lakes, lagoon, etc.) and, hence reduce the photosynthetic activity of aquatic flora. As a result, there is a significant reduction in the oxygenation capacity of the water, disturbing the whole aquatic ecosystem and the food chain (Xu et al. 2005; Merzouk et al. 2010; Valh et al. 2011; Chhabra et al. 2015). Dissolved dyes can also cause an increase in air pollution in terms of stringent foul odors, inorganic carbon deposits, changes of the soil matrix (Senthilkumar et al. 2011) and ground water systems are affected due to leaching through the soil (Khaled et al. 2009).

When dyeing wastewater is disposed of without appropriate, or any, treatment, into water bodies, it can form a thin layer of discharged dyes over the surface and also decrease the amount of dissolved oxygen in the water, thereby affecting the aquatic fauna (Ali 2010) and increasing the BOD of the contaminated water

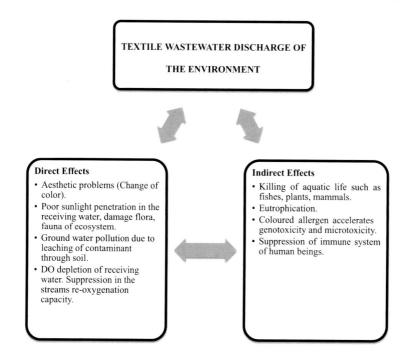

Fig. 4 Effects of textile wastewater into the environment. Reprinted with permission of Verma et al. (2012). Copyright © 2011 Elsevier Ltd.

(Annuar et al. 2009). Apart from this, several dyes and their decomposition derivatives have proved to be toxic to aquatic life such as crustaceans (Bae and Freeman 2007; Ferraz et al. 2013), microorganisms (Gottlieb et al. 2003), fish (Oliveira et al. 2016), and algae (Sponza 2006).

4 Coagulation/Flocculation Process (CF)

Currently, there is an increased interest in the decolorization and decontamination of industrial textile wastewater (Garcia et al. 2009). Different treatment technologies have been studied in order to solve the problems caused by the toxic substances contained in industrial textile wastewater, such as electrocoagulation (Naje et al. 2016; Ozyonar 2015), adsorption (Simionato et al. 2014; Góes et al. 2016), photocatalytic process (Souza et al. 2016, 2017; Garcia et al. 2012; Palácio et al. 2012), ozonation (Castro et al. 2016), membrane bioreactor (Jegatheesan et al. 2016; Deveci et al. 2016) and anaerobic/aerobic biological treatment (Abiri et al. 2016). However, these methods are neither economically nor technologically suitable for large scale use and normally require the combination of two or three methods to achieve an appropriate level of color removal (Robinson et al. 2001; de Souza et al. 2016).

Coagulation/flocculation (CF) is the most widely used technique in industrial wastewater treatment worldwide (de Souza et al. 2016) because it is relatively simple and inexpensive to operate (Fu and Wang 2011). It has been reported that the Egyptians were using aluminum sulfate (alum) to cause suspended particles to settle in water as early as 1500 BC. Although the early Romans were also familiar with alum, it was not until 77 AD that its utilization as a coagulant in water treatment was firs recorded (Teh et al. 2016). Today, the coagulation-flocculation process is a vital step in removal of colloidal particles, natural organic matter, microorganisms and inorganic ions present in untreated water (Eckenfelder 1966; Renault et al. 2009; Kakoi et al. 2016).

The inherent disadvantage to this process is its generation of large quantities of chemical sludge that is classified as hazardous waste that must be disposed of in secure landfills. The process also increases the total dissolved salt content in the treated wastewater, increasing desalination costs by a substantial margin. On the other hand, coagulant aids are inorganic materials that, when used alongside a main coagulant improve or accelerate the process of coagulation and flocculation by producing quick forming, dense and rapidly settling flocs. Coagulant aids increase the density of slow settling flocs and add toughness to the flocs so that they do not break up during the mixing and settling processes (Holt et al. 2002; Sahu and Chaudhari 2013).

The efficiency of CF its operational costs depend on several factors, including the coagulant type and dosage, mixing conditions, pH, temperature, ionic strength, as well as the nature and concentration of the organic matter, the total dissolved solids, the size and distribution of the colloidal particles in suspension among others (Rodrigues et al. 2008; Santo et al. 2012).

4.1 Physical-Chemical Concepts About Coagulation/Flocculation Process

Colloids that contain charged particles remain stable, since the forces of Coulomb repulsion between charges of the same sign are stronger than attractive van der Waals forces. For coagulation to occur it is necessary to destabilize these electric charges. One of the most used models for the explanation of the coagulation phenomenon is the double diffuse layer model.

According to the dual layer model, negatively charged micelles are associated with an electric field around them in the water that contains positively charged (+) cations and negative (−) anions. Immediately, the hydrated cations will be attracted and will be disposed on the negative surface of the micelles to neutralize the electric field, obeying Coulomb's law given by Eq. (1).

$$F = \frac{q^+ q^- (micelle)}{4\Pi\varepsilon_0 r^2} \qquad (1)$$

However, as the electric field is neutralized by the positive charges, there remains an electric field "residue", whose force is counterbalanced by the kinetic energy (temperature dependent) of the ions, and a disorganized (or unstructured) region is formed because the force resulting from the kinetic energy of the particles present on it is greater than or equal to the force of attraction resulting from the electrostatic and van der Waals forces.

The diffuse layer theory has as basic postulates:

I. Interactions are electrostatic;
II. The retention force decreases between the negative charges (−) of the micelle and the adsorbed cations with distance, establishing two defined attraction zones: adsorbed cations—"internal solution"; cations and anions in equilibrium—"external solution";
III. A balance governed by thermodynamics is established between the two: the increase of a factor in one of the phases implies compensation in the other and vice versa.

There are two double layer diffusion models:

In the Helmhorltz model (originating in mid-1879), the counter ions of the solution move to the vicinity of the charged micelle, forming a layer of ions adhering to the micelle and creating a potential Ψ there. The Helmhorltz model assumes that the decrease in the force of attraction between counterions and micelles, that is, the decrease in potential, is rapid with distance.

According to the diffuse double layer model (Stern Model), there is an adhered layer and, outside of it, a diffuse layer around the colloidal micelle. Therefore, the potential Ψ and the attractive forces between the micelle and the counterions first fall rapidly at first (in the adhered layer), and then slowly in the diffuse layer, as shown in Fig. 5.

Analyzing the two models for the diffuse double layer it can be concluded that there are three potentials that can be estimated:

The Ψo potential on the micelle surface.

The potential Ψ on the inner surface of the double layer where the diffuse layer begins.

The potential ς (zeta) in the shear plane. The shear plane is the section of the double layer that moves along with the colloidal particle. It is located between the inner and outer surfaces of the double layer. Since the particle cannot be separated from its counterions, the only potential that can be determined is the potential ς, and the knowledge of this potential is important in order to achieve destabilization of the colloidal particles.

The resultant interaction between the repulsive Coulomb forces and attractive van der Waals forces creates an energy barrier that prevents the approach of colloidal particles, since the repulsive forces are more intense. This energy barrier

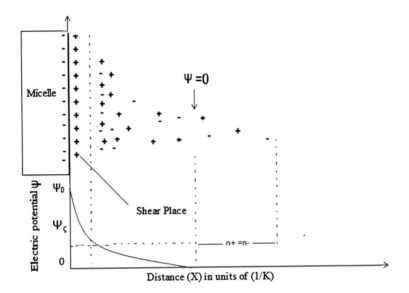

Fig. 5 The double layer model proposed by Stern. Adapted by Lenzi et al. (2009)

varies with pH (pH dependent charge). The distance between micelles at which flocculation can occur, which is achieved when the potential ç is null can be approximated; this point is called the isoelectric point. The isoelectric point can be achieved by neutralizing the micelle charges or by decreasing the thickness of the double electrical layer.

From an electrostatic point of view, coagulation is the reduction of the zeta potential by the addition of specific ions. Coagulation occurs when the addition of a cationic electrolyte decreases the zeta potential, because the electrolyte reduces repulsive forces and allows the action of van der Waals attractive forces to promote agglutination. The required electrolyte dosage depends on the concentration of the colloids (Leme 1982; Lenzi et al. 2009).

In order for efficient coagulation to take place, intense mixing must be achieved by suitable agitation, which has the function of producing turbulence and is characterized by the velocity gradients it creates. An intense mixture ensures uniform distribution of the coagulant into the water by contacting the particles in suspension before the reaction is complete. Flocculation consists of clustering and compaction of the suspended particles in large sets called flakes, which is achieved by limited slow stirring to avoid breaking already formed flakes.

In the mixture, flotation decisively influences the preparation for decanting. It refers to the gathering of coagulated material into larger and denser particles, called flakes (Leme 1979).

4.2 CF Mechanism

The process of coagulation or destabilization of colloidal particles may occur through four types of mechanisms: adsorption and bridging; double-layer compression; charge neutralization; and sweep coagulation. Each of these mechanisms is considered below.

4.2.1 Adsorption and Bridging

Bridging between particles occurs with the introduction of long-chain polymers or polyelectrolytes, as these coagulants are capable of extending into the solution to capture and bind multiple colloids together. The bridging efficiency is further improved when coagulants with larger molecular weights are used, due to their extended polymeric chains. Natural polymers such as polysaccharides and proteins can also induce coagulation via bridging. Such a mechanism has been recognized since the 1950s (Ruehrwein and Ward 1952) and is extremely important in practice. An essential requirement for bridging flocculation is that there should be sufficient unoccupied surface on the particles for attachment of segments of polymer chains adsorbed on other particles. It follows that the adsorbed amount should not be too high, otherwise the particle surfaces will become so highly covered that there are insufficient adsorption sites available (Bolto and Gregory 2007; Choy et al. 2015).

In general, if the molecular weight is high and the charge density is low, the polymer adsorbs on the particle surface in such a way that tails and loops extend far beyond the surface and can interact with other particles (Rasteiro et al. 2008). The consequence of bridging flocculation is that the flocs produced can be much stronger than those formed when particles are destabilized by simple salts (Li et al. 2006).

4.2.2 Double-Layer Compression

This mechanism involves the reduction of the double layer around the colloidal particle by a change in ionic strength induced by the addition of a different electrolyte, which results in the destabilization of colloid (Teh et al. 2016). Under stable conditions in which the concentration of counter-ions is low, colloidal particles are unable to get close to each other because of their thick double electrical layer. However, as the concentration of counter-ions increases via the addition of salts, the diffuse layer becomes thinner and particles can approach more closely before experiencing repulsion. It has been speculated that compression of the electrical double layer is the dominant coagulation mechanism for divalent ions, principally Ca^{2+} and Mg^{2+}, in the normal pH range of water treatments (Duan et al. 2009).

4.2.3 Charge Neutralization

Charge neutralization is the process of adding cationic metals or polymers to neutralize the negative charges of the particles. This coagulation mechanism involves the adsorption of an oppositely charged coagulant on the colloidal surface. It is well known that electrostatic interactions give strong adsorption in these systems and that neutralization of the particle surface and even charge reversal can occur. There is thus the possibility that flocculation could occur simply as a result of the reduced surface charge of the particles and hence a decreased electrical repulsion between them (Bolto and Gregory 2007; Choy et al. 2015; Teh et al. 2016).

4.2.4 Sweep Coagulation

Sweep coagulation is a mechanism related to metal coagulants like Al^{3+} and Fe^{3+}. If a high concentration of these metals is added to water, a large quantity of metal hydroxide will be produced. This amorphous hydroxide will settle, and sweep colloidal particles in its way downward (Renault et al. 2009).

Sweep flocculation generally gives considerably better particle removal than destabilization by charge neutralization alone. At least part of the reason is the greatly improved rate of aggregation, because of the increased solid concentration. Hydroxide precipitates tend to have a rather open structure, so that even a small mass can give a large effective volume concentration and, hence, a high probability of capturing other particles (Duan et al. 2009).

5 Chemical Coagulants

The CF process is a complex phenomenon that involves several interrelated parameters, thus it is very important to choose the right coagulant. Chemical coagulants can be categorized into three groups, as shown in Fig. 6 (Verma et al. 2012).

The most common inorganic coagulants are aluminum and iron salts such as $AlCl_3$, $Al_2(SO_4)_3$, $FeCl_3$, and $Fe_2(SO_4)_3$ due to their low cost, ease of use, handling, storage and mixing properties (Matilainen et al. 2010; Sher et al. 2013). When added to water, aqueous Al (III) and Fe (III) salts dissociate to their respective trivalent ions, i.e. Al^{3+} and Fe^{3+}. After that they are hydrolyzed and form several soluble complexes possessing highly positive charges, thus adsorbing onto the surfaces of negative colloids (Leme 1979; Teh et al. 2016).

Despite the wide applicability of these salts, their use has been questioned due to the iron salts producing brownish coloring of equipment, and environmental impacts caused by high concentrations of residual aluminum in treated wastewater and/or sludge (Matilainen et al. 2010; Zhu et al. 2011; de Souza et al. 2016a). Furthermore, epidemiological, neuropathological and biochemical studies suggest a

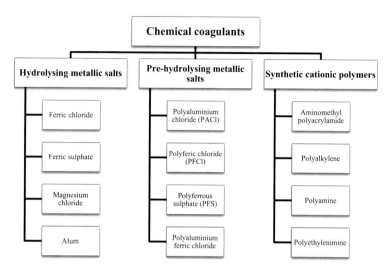

Fig. 6 Schematic category of chemical coagulants by their efficiency. Reprinted with permission of Verma et al. (2012). Copyright © 2011 Elsevier Ltd.

possible link between the neurotoxicity of aluminum and the pathogenesis of Alzheimer's disease (McLachlan 1995; Polizzi et al. 2002; Banks et al. 2006; Walton 2013).

Textile wastewater treatment using inorganic coagulants is not always efficient enough. There are several drawbacks associated with their usage, such as ineffectiveness at low temperatures, harmful effects on human health, the production of large sludge volumes and changes in the pH of the treated water (Choy et al. 2014; Shamsnejati et al. 2015). Under different environmental conditions such as extreme pH and very low or very high temperatures, very sensitive, fragile flocs may be produced, which result in poor sedimentation. These flocs may rupture under any type of physical force (Verma et al. 2012).

To improve the efficiency of the coagulation process, to minimize the hazards of the use of inorganic salts and to obtain wastewater of good quality and fast sedimentation, organic polymers may be recommended (Aguilar et al. 2005; Verma et al. 2012; de Souza et al. 2016a). Organic polymers are water soluble and can be classified into two broad categories: natural and synthetic (Zahrim et al. 2011). Coagulants based on synthetic polymers contain pollutants that may threaten health and the environment. Some of their derivatives are non-biodegradable and the intermediate products of their degradation are hazardous (Oladoja 2015). These pollutants are derived from unreacted monomers (such as acrylamide, diallymethyl ammonium chloride), unreacted chemicals used to produce the synthetic polymers (such as epichlorohydrin, formaldehyde), and reaction by products of the polymers in water (such as N-nitrosodimethylamine). The formation of undesirable secondary products may also occur, for example, acrylamide is very much toxic, gives severe neurotoxic effects and has been shown to be carcinogenic, particularly affecting the thyroid, mammary and adrenal glands, the scrotum and the oral cavity (Bratby 2007; Zahrim et al. 2011). Synthetic polyacrylamides are the main organic polymer used in water and waste treatment plants, but in many countries their use has already been stopped on the recommendation of the World Health Organization (WHO 2003; Beltrán-Heredia et al. 2011). Hence, many natural coagulants/flocculants have instead been used for the treatment of textile wastewaters (Teh et al. 2016).

6   Natural Coagulants—Plant-Based Coagulants

In view of the need to overcome the drawbacks of inorganic coagulants and synthetic polymers associated with growing environmental concerns worldwide, there is a need to consider other potential alternatives for textile wastewater treatment in order to minimize environmental damage and improve the wellbeing of human populations. Therefore, researchers have shown significantly more interest in the development of natural polymers as coagulants in recent years (Choy et al. 2014, 2016; de Souza et al. 2014, 2016a; Freitas et al. 2015; Shamsnejati et al. 2015).

The use of natural coagulants for the clarification of water and wastewater has been recorded throughout human history since ancient times and it is still current today (Sanghi et al. 2002). Natural organic polymers have been used for more than 2000 years in India, Africa, and China as effective coagulants and coagulant aids at high water turbidities (Asrafuzzaman et al. 2011).

Natural coagulants can be divided into cationic, anionic or nonionic, and hence are also termed as polyelectrolytes (Yin 2010). Examples of natural cationic polymers are chitosan and cationic starches, while some examples of anionic polymers are sulfated polysaccharides and modified lignin sulfonates. Starch and cellulose derivatives are examples of non-ionic natural polymers (Renault et al. 2009). Particles can aggregate and settle out of solution through four basic mechanisms: double layer compression; sweep flocculation; adsorption and charge neutralization; and adsorption and interparticle bridging. Natural coagulants generally exhibit two types of mechanism: adsorption and charge neutralization; and adsorption and interparticle bridging. Adsorption and charge neutralization can occur when suspended particles in solution sorb to oppositely charged ions, while interparticle bridging occurs when polysaccharide chains of coagulants can attach to multiple particles so that particles are bound to the coagulant and need not contact one another. The existence of adsorption and interparticle bridging between dye molecules and polysaccharides is due to the interaction of π-electron systems of dyes and OH^- groups of polysaccharides (Fig. 7) (Miller et al. 2008; Yin 2010; Verma et al. 2012).

Based on their origin of production, natural coagulants can be divided into three categories; plant, microorganism or animal-based as shown in Fig. 8. However, available sources of plant-based coagulants are much more widespread than

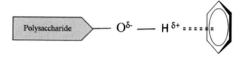

Fig. 7 Schematic representation of intermolecular interaction between π-electron from dye molecule and hydroxyl group of polysaccharide. Reprinted with permission of Verma et al. (2012). Copyright © 2011 Elsevier Ltd.

Fig. 8 Schematic categorization of natural coagulants

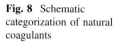

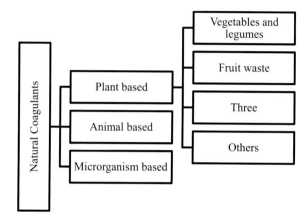

animal-based coagulants, thus plant-based coagulants could be potential alternatives to chemical coagulants and have gradually gained in importance over the years (Choy et al. 2015).

6.1 Advantage and Disadvantage of Using Plant-Based Coagulants

Organic polymeric compounds have advantages over inorganic materials, as they possess several novel characteristics such as their ability to produce large, dense, compact flocs that are stronger and have good settling characteristics. Natural polymers may potentially be applied not only in food and fermentation processes, and in downstream processing, but also in water and wastewater treatment (Renault et al. 2009) and they can be applied in textile wastewater treatment.

The advantages of using plant-based coagulants in place of inorganic coagulants and/or organic polymers include lower coagulant dose requirements, smaller increases in the ionic load of the treated wastewater, and reduced levels of metals in the treated wastewater (Kakoi et al. 2016). Due to their low toxicological risk (generally free of toxicity) and high biodegradability, they are safe to human health and aquatic life, consequently having a smaller environmental impact (Choy et al. 2016;

de Souza et al. 2016a). One obvious advantage of using renewable materials is their minimal net effect on global warming (Sharma et al. 2006). There are several other advantages: renewable materials generally do not produce secondary pollution (Renault et al. 2009); they usually have a large number of surface charges, which increases the efficiency of the coagulation process; they show environmentally friendly behavior (Freitas et al. 2015); they do not alter the pH of the water under treatment (Oladoja 2015) they are unlikely to produce treated water with extreme pH (Yin 2010); they are also non-corrosive, which eliminates concerns about pipe erosion (Choy et al. 2014).

Plant-based coagulants have been found to generate not only much smaller sludge volumes (up to five times lower) (Ndabigengesere et al. 1995) but also sludge of higher nutritional value (Choy et al. 2014). Wastewaters treated with them pose no risk to biological organisms, in contrast to synthetic coagulants. Furthermore, the sludge produced by natural polymers can be treated by biological processes or disposed of as a soil conditioner due to its non-toxicity (de Souza et al. 2014). As such, sludge treatment and handling costs are lowered, making plant-based coagulants a more sustainable option (Choy et al. 2016).

Using plant-based coagulants can reduce wastewater treatment costs and these advantages are especially augmented if the plants are grown, cropped, and processed locally (Sanghi et al. 2006a; de Souza et al. 2014; Freitas et al. 2015). The raw plant extracts are often available locally and hence, the natural plant-based coagulants are more cost-effective than imported chemicals. Since natural coagulants do not consume alkalinity unlike alum, pH adjustments can be omitted and this provides extra cost savings (Choy et al. 2014; Oladoja 2015).

Plant-based polymeric coagulants are considered "green" because they satisfy the requirements of green technology such as sustainability—meeting the needs of society in ways that can continue into the future without damaging or depleting natural resources; cradle to cradle design—creating products that can be fully reclaimed or re-used; source reduction—reducing waste and pollution by changing patterns of production and consumption; innovation—developing alternatives to technologies that have been demonstrated to damage health and the environment; and viability—creating a center of economic activity around technologies and products that benefit the environment (Oladoja 2015). In the age of climate change, depletion of the earth's natural resources, and widespread environmental degradation, application of these coagulants in textile wastewater technologies is a vital effort and can contribute to advancing global sustainable development initiatives (Miller et al. 2008; Yin 2010). Figure 9 summarizes the benefits of using natural coagulants as an alternative to chemical coagulants.

Plant-based polymeric coagulants also have some disadvantages. Natural polymers are generally non-toxic but synthetic polymers are more effective due to the possibility of controlling their properties such as the number and type of charged units and the molecular weight. Moreover, due to their biodegradability, natural polymers have a shorter storage life than synthetic polymers (Zahrim et al. 2011). Despite the promise shown by these materials, their application has not yet made

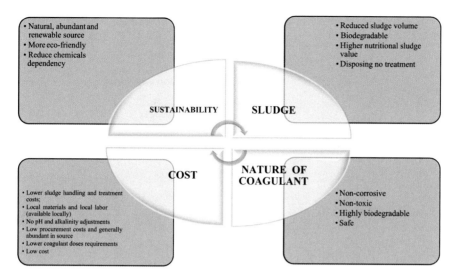

Fig. 9 Advantages of natural coagulants over chemical coagulants. Reprinted with modifications and permission of Choy et al. (2014). Copyright © 2014. Published by Elsevier B.V.

the transition from bench scale to field trials and application in industrial treatment plants (Oladoja 2015).

One of the problems in the use of plant-based coagulants is the substantial increase in the organic load of the treated water, which may result in undesired and increased microbial activities, having serious implications on subsequent disinfection processes using chlorine. Organic matter is regarded as the source of odor, color, and taste (Choy et al. 2014; Oladoja 2015). These organic matters might act a precursor of trihalomethanes (THMs), of which many are carcinogenic and also require chlorine treatment (Anastasakis et al. 2009).

7   Plant-Based Coagulants on Textile Wastewater Treatment

In recent years, natural coagulants extracted from plants have been successfully used in textile wastewater treatment and the number of publications in this area has been growing significantly. This is due to advantages such as biodegradability, safety to human health, low cost, low toxicity and eco-friendliness and abundance. Table 6 summarizes the use of plant-based coagulants in textile wastewater treatment.

Table 6 Research summaries of textile wastewater treatment by coagulation/flocculation process using plant-based coagulants

Coagulant	Sample	Efficiency of coagulation/flocculation process			References
		Turbidity removal (%)	COD removal (%)	Color removal (%)	
Acanthocereus tetragonus	Congo Red			96.7	Chethana et al. (2015)
Alginate from *Sargassum* sp. (Brown Algae)	Sulphur Black Dye			98.2	Vijayaraghavan and Shanthakumar (2015)
Anacardium occidentale	Congo Red Methyl Orange				Edilson et al. (2016)
Azadirachta indica Cicer arietinum Acanthocereus tetragonus Moringa oleifera	Congo Red			95 (Ai) 89 (Ca) 97 (At) 99 (Mo)	Chethana et al. (2016)
Cactus *Cereus peruvianus*	RTW (DF)	85.4	58.3		De Souza et al. (2016a)
Cactus *Opuntia ficus-indica*	RTW (LJ And DF)	91.3 (LJ) 93.7 (DF)	64.5 (LJ) 87.3 (DF)		De Souza et al. (2014)
Cactus *Opuntia ficus-indica*	Painting wastewater	82.6	78.2	88.4	Vishali and Karthikeyan (2015)
Cassia angustifolia seed	Mixture Dyes: Acid Sendula Red Direct Kahi Green Reactive Remazol Brilliant Violet			84 (ASR), 83 (DKG) and 31 (RRBV)	Sanghi et al. (2002)
Cassia fistula Linn seed	Reactive Blue 19 Reactive Black 5			62.0 (RB19) 55.7 (RB5)	Perng and Bui (2015)
Cassia fistula seed	Reactive Red 195			57.8	Manh et al. (2015)
Cassia fistula seed	RTW (DF and DFI)			37.5 (DF) and 71.3 (DFI)	Hanif et al. (2008)

(continued)

Table 6 (continued)

Coagulant	Sample	Efficiency of coagulation/flocculation process			References
		Turbidity removal (%)	COD removal (%)	Color removal (%)	
Cassia javahikai seed	Direct Bordeaux BW Direct Orange Acid Sandolan Red Reactive Procion Brilliant Blue RS RTW			58 (DB), 63 (DO), 5 (ASR), 8 (RPBB) and 25 (RTW)	Sanghi et al. (2006b)
Descurainia sophia L.	Neutral Red Dye			90.2	Ahmadi et al. (2016)
Enteromorpha	Reactive Blue 14 Dye			85–88	Zhao et al. (2014)
Furcraea sp.	Textile industry wastewater	93	80	89	Lozano-Rivas et al. (2016)
Grape (*Vitis vinífera*) *seed*	Malachite Green Crystal Violet			60 (MG) and 80 (CV)	Jeon et al. (2009)
Moringa oleifera Phaseolus vulgaris	Congo Red			83 73	Vijayaraghavan and Shanthakumar (2015)
Moringa oleifera seed	Direct Black 19 Dye			94.2	Tie et al. (2015)
Moringa oleifera seed	Alizarin Violet 3R			95	Beltrán-Heredia et al. (2009a)
Moringa oleifera seed	Chicago Sky Blue 6b			99	Beltrán-Heredia and Martín (2008)
Moringa oleifera seed	RTW	92.3	92.0	98.8	Hemapriya et al. (2015)
Moringa oleifera seed	Carmine Indigo			80	Beltrán-Heredia et al. (2009b)
Moringa oleifera seed	Reactive Yellow		62	90	Veeramalini et al. (2012)
Ocimum basilicum seed	Congo Red RTW (DF)		61.6 (Dye) 80.4 (RTW)	68.5 (Dye) 76.4 (RTW)	Shamsnejati et al. (2015)

(continued)

Table 6 (continued)

Coagulant	Sample	Efficiency of coagulation/flocculation process			References
		Turbidity removal (%)	COD removal (%)	Color removal (%)	
Okra waste (A. esculentus)	RTW (JL)	97.2	85.7	93.6	Freitas et al. (2015)
Plantago major L.	Neutral Red		81.6	92.4	Chaibakhsh et al. (2014)
Plantago ovata seed	RTW (TFI)		89.6		Ramavandi and Farjadfard (2014)
Plantago psyllium	Golden Yellow Dye Reactive Black 5			71.4 (GY) 35 (RB5)	Mishra and Bajpai (2005)
Plantago psyllium	RTW	90 (SS)			Mishra et al. (2002)
Portunus sanguinolentus	Paint effluent	66.85	59	92.29	Vishali et al. (2016)
Surjana seed powder and Maize seed powder	Congo Red dye			98.0 (SSP) and 89.4 (MSP)	Patel and Vashi (2012)
Tamarindus indica	Golden Yellow Direct Fast Scarlet			60 (GY) 25 (DFS)	Mishra and Bajpai (2006)
Vegetal tannin (Polysep3000)	RTW		40–50	96	Aboulhassan et al. (2005)
Vegetal tannin (Tanfloc®)	RTW (JL)	92,1	49.2	76.1	De Souza et al. (2016b)

ASR: Acid Sendula Red Dye; Ai: *Azadirachta indica*; At: *Acanthocereus tetragonus*; Ca: *Cicer arietinum*; COD: chemical oxygen demand; CV: Crystal Violet Dye; DF: dyeing fabric; DB: Direct Bordeaux BW Dye; DFS: Direct Fast Scarlet Dye; DKG: Direct Kahi Green Dye; DO: Direct Orange Dye; GY: Golden Yellow Dye; LJ: laundry jeans fabric; Mo: *Moringa oleifera*; MSP Maize seed powder; MG: Malachite Green Dye; RB5: Reactive Black 5 Dye; RB19: Reactive Blue 19 Dye; RPBB: Reactive Procion Brilliant Blue RS; RRBV: Reactive Remazol Brilliant Violet Dye; RTW: real textile wastewater; SS: suspended solids; SSP: Surjana seed powder; TFI: textile finishing industry

7.1 Tannin

Tannin is a general name given to large polyphenol compounds, it is water-soluble polyphenolic anionic polyelectrolyte; with a molecular weight around 500 thousand. They are obtained from natural materials, for example, the organic extract from bark and wood of trees such as Acacia, Castanea, or Schinopsis (Yin 2010).

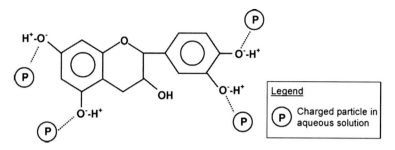

Fig. 10 Schematic representation of basic tannin structure in aqueous solution and possible molecular interactions. Reprinted with permission of Yin (2010). Copyright © 2010 Elsevier Ltd.

They also is found in some plants, such as in the seeds of *Cassia obtusifolia* (Subramonian et al. 2014) and *Castanea sativa* trees (Chestnut) (Oladoja 2016). The chemical structure is complex and cannot be accurately determined, but it is known that carboxyl and hydroxyl groups predominate. Studies have shown the efficiency of tannin as a coagulant and flocculant (Choy et al. 2014; Oladoja 2016). Widely known and studied, tannin may be used as coagulant in the treatment of textile wastewater, and can be modified for use as a cationic polyelectrolyte to remove anionic compounds (Beltrán-Heredia et al. 2011).

Figure 10 illustrates the schematic representation of basic tannin structure in aqueous solution and possible molecular interactions that induce coagulation. It is common knowledge that phenolic groups can easily deprotonate to form phenoxide which is stabilized via resonance. This deprotonation is attributed to delocalization of electrons within the aromatic ring which increases the electron density of the oxygen atom (Yin 2010).

7.2 *Moringa oleifera*

Moringa oleifera is a tropical plant found in Asia, India, Africa and Latin America, is reported that its seeds are used to clean turbid river water in rural communities in African countries. It is the most studied natural coagulant in environmental science. Its coagulant potential can be improved by the addition of bivalent cations as Ca^{2+} and Mg^{2+} (Yin 2010). It also can be used to improve the efficiency of other natural coagulants.

Kansal and Kumari (2014) reported the coagulant efficacy of *Moringa oleifera* on seven pigments, with a removal of 60–99%. When applied as coagulant in Congo Red pigment solution, a concentration of only 25 mg/L of *M. oleifera* seeds solution removed 98% of the pigment. The seed also showed satisfactory removal of turbidity caused by the pigment Black 19, demonstrating that this species is one of the most promising natural coagulant species for textile wastewaters (Tie et al. 2015).

7.3 Ocimum basilicum

Ocimum basilicum is popularly known as basil, and can be found in tropical regions in Asia, Africa, Central America and South America. It has many different applications, from use as a condiment to applications in traditional medicine. Its seeds have been used in CF processes and have shown good results. When soaked in water, the seeds create mucilage rich in pectin. The seed extract is rich in polysaccharides, glucomannan and $1 \rightarrow 4$ bounded xylan, with a small proportion of glucan. After polysaccharide extraction in water, a viscous solution was obtained, which was used as coagulant to treat textile wastewater and Congo red pigment solution. Using only 1.6 mg/L of coagulant, it was possible to remove 61.6% of the COD and 68.5% of the color from a 50 mg/L solution of pigment. When used on real textile wastewater, the removal was of 50.9% and 56.6% for color and COD, respectively, using 9.6 mg/L of *Ocimum basilicum* coagulant solution (Shamsnejati et al. 2015; Oladoja 2015).

7.4 Cereus peruvianus

There are many studies with different cactus species describing their use as coagulants and flocculants (Zhang et al. 2006; Miller et al. 2008). De Souza et al. (2016) reported the efficiency of *Cereus peruvianus* for the treatment of textile wastewater. Mucilage extracts were applied to tests of CF. ANOVA analysis was performed to measure the importance of some significant factors (pH, mucilage and inorganic coagulant). The data were shown to fit a quadratic model and the mucilage concentration was significant for the removal of turbidity and COD. The best conditions were: inorganic coagulant 320.0 mg/L, mucilage 10.0 mg/L and pH 5.0. Removal of turbidity and COD of $85.4\% \pm 0.5$ and 58.3 ± 0.2, respectively, were achieved.

7.5 Galacturonic Acid

Studies performed with different plant species have reported the presence of galacturonic acid, and attributed efficiency in the CF process to its presence. Freitas et al. (2015) reported the presence of galacturonic acid in okra (*A. esculentus*) where it is probably present in the polymer form, polygalacturonic acid, allowing it to provide a "bridge" to adsorb particles by van der Waals and Coulomb interactions. Figure 11 shows how polygalacturonic acid can interact in the CF process.

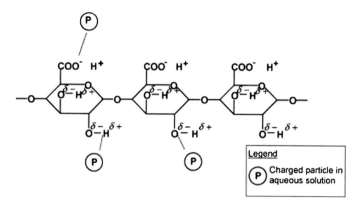

Fig. 11 Schematic representation of polygalacturonic acid in aqueous solution and possible dominant molecular interactions associated with adsorption and bridging. Reprinted with permission of Yin (2010). Copyright © 2010 Elsevier Ltd.

7.6 Cassia *Seed Gums Based Coagulants/Flocculants*

Cassia plants are a known source of seed gums which are usually galactomannans that have close structural resemblance to many commercial seed gums, such as guar and locust bean gums. They are considered as non-conventional renewable reservoirs for galactomannan seed gums. Thus, the properties of Cassia seed gums in general can be tailored by chemical modification whereupon they can be exploited as useful dye flocculants and heavy metal adsorbents depending upon their solubility in water. Though galactomannans from Cassia seeds are nonionic polysaccharides, their adsorption performance is comparable to that of chitin and chitosan, and superior to other polysaccharides (Kalia and Averous 2011).

7.6.1 Cassia Javahikai

Sanghi et al. (2006b) used a *Cassia javahikai* seed gum as a coagulant for the decolorization of textile dye solutions, in combination with very low doses of polyaluminium chloride (PAC). Graft copolymerization with acrylamide was performed to favorably modify the properties of the seed gum. C. *javahikai* (CJ) seed gum, and its copolymer grafted with acrylamide were synthesized in the presence of oxygen using a potassium persulfate/ascorbic acid redox system. Seed gum CJ and its grafted polymer C. *javahikai* (CJG) proved to be very efficient coagulants for the decolorization of direct dyes. In combination with PAC, CJ and CJG were found to be efficient coagulant aids for acids as well as for Procion dyes. The performance of CJG over CJ was more enhanced for direct dyes. Being biodegradable and safe to human health, they can be potential alternatives to conventionally used chemical coagulants.

7.6.2 Cassia angustifolia

Sanghi et al. (2002) evaluated naturally occurring *Cassia angustifolia* (CA) seed gum against the chemical coagulant polyaluminium chloride (PAC) for its ability to remove color from synthetic dye solutions. Three groups of dyes: Acid Sendula Red, Direct Kahi Green and Reactive Remazol Brilliant Violet were chosen for the case study. CA was found to be a good working substitute alone or in conjunction with a very low dose of PAC for decolorization of acid and for direct, but not reactive, dye solutions. It may be concluded that Cassia seed gum can be an effective coagulant aid for direct and acid dyes. It can act as a working substitute, partially or fully, for synthetic chemical coagulants such as PAC.

7.6.3 Cassia fistula

Hanif et al. (2008) carried out textile wastewater treatment using a *Cassia fistula*-based natural coagulant. The dosage of *C. fistula* used in the study was in the range 250–1500 mg L^{-1}. The reduction of TDS was of 86.02–88.39%, and the COD presented a reduction of 37.45% at a dosage of 1500 mg L^{-1}. The efficiency of treatment of textile industry wastewater was found to be dependent on the dosage of the coagulant as well as on the pH of the wastewater.

Perng and Bui (2015) developed a study in order find a new environmentally friendly coagulants that could partially replace conventional polyaluminum chloride (PAC), which has been shown to be toxic to aquatic environments. Gum extracted from the seeds of Cassia *fistula* Linn. (CF) was investigated for the decolorization of the reactive dyes Blue 19 (RB19) and Black 5 (RB5), using jar-test experiments. The decolorization of these dyes by CF gum was highly dependent on pH, probably because it affects the structure of the dyes, the gum and their interactions. The best results were obtained at pH 10. When the gum alone was used for RB19 and RB5 removal, the best treatment efficiencies achieved were 62.0% and 55.7%, respectively. However, a combination of low dosages of gum (80 and 120 mg L^{-1}) and PAC (132 mg L^{-1}) showed rather high removal efficiencies of greater than 92%. The amount of PAC coagulant used was reduced by about 40%. These results indicated the potential of using CF gum as a "green" coagulant or for color removal in textile wastewater.

Manh et al. (2015) studied the ability of *Cassia fistula* seed gum to remove the color from Reactive Red 195 solutions in Jar-test experiments. This low-cost coagulant showed great potential for decolorization of Reactive Red 195 dye solution. The best color removal performance obtained was 57.8%, using 200 mg/L of gum, an agitation speed of 45 rpm, IDC at 10 mg/L and 60 min of treatment at pH 10. Decolorization of Reactive Red 195 dye solutions by coagulation with the gum was highly dependent on the pH and coagulant dosage.

7.6.4   Cassia obtusifolia

Subramonian et al. (2014) investigated the potential use of natural *Cassia obtusifolia* seed gum in the treatment of raw and undiluted PPME through coagulation. The recommended conditions (initial pH 5, dosage of 0.75 g/L, 10 rpm and 10 min slow mixing, with 1 min settling time) allowed removal of total suspended solids and chemical oxygen demand of up to 86.9 and 36.2%, respectively. The findings from the present study showed that the coagulation efficiency using *C. obtusifolia* gum was comparable to that of alum. Characterization of the process showed that *C. obtusifolia* gum, alum and their flocs all possessed distinctive features. The study demonstrated that *C. obtusifolia* gum is a promising and effective natural coagulant that can substitute for harmful inorganic coagulants such as alum in the treatment of industrial effluents.

7.6.5   Cassia tora

Sharma et al. (2006) carried out a chemical modification of *Cassia tora* and guar gum through different substitution and grafting reactions. The modified products were tested against kaolin under laboratory conditions vis-a-vis a polyacrylamide-based synthetic flocculant. The tests were performed with the following modified gums: CB-CTG-carbamoylethyl *C. tora* gum; CE-CTG-cyanoethyl *C. tora* gum; CM-CTG-carboxymethyl *C. tora* gum; Q-CTG-quaternized *C. tora* gum; CTG-g-AA-acrylamide-grafted *C. tora* gum; CTG-g-A-acrylonitrile-grafted *C. tora* gum; CB-GG-carbamoylethyl guar gum; GG-g-MMA-methyl methacrylate-grafted guar gum.

The flocculation efficiency of all the chemically modified products was studied. In the case of products synthesized by substitution reactions the flocculation was greatest with Q-CTG (99.65%) due to its effective charge neutralization as well as bridging, attributed to the cationic functional groups (quaternary ammonium groups) introduced into the backbone. In the case of CB-CTG and CB-GG, the settling of particles was 99.05% and 99.0%, respectively attributed to the carbamoyl-ethyl groups introduced into the biopolymer. Settling of particles was 98.58% with CE-CTG due to the introduced cyanoethyl groups. The settling of particles was 97.80% for CM-CTG because it has a large hydrodynamic volume compared with the other substituted products, due to the presence of more hydrophilic carboxymethyl groups in its backbone.

It can be concluded that the flocculation efficiency of most of the chemically modified products of *C. tora* and guar gum is better than that of synthetic flocculants. These modified products can be further exploited for the treatment of many industrial effluents.

7.6.6 Galactomannans of Plants of the Genus *Cassia*

Galactomannan is a polysaccharide and it is present in most legumes. They constitute the second largest group of polysaccharide reservoirs in the plant world, and can be found in several mucilaginous plants. They are commonly found in the seed endosperm of legumes (Reid 1985; Scherbukhin and Anulov 1999). This class is artificially defined as containing more than 90% mannose, forming a β-$(1 \rightarrow 4)$ linear chain without ramification. They may or may not be branched with galactose with α-$(1 \rightarrow 6)$ bonds (Reid and Edwards 1995).

They are present in 13 vegetable families: *Annonaceae, Compositae, Convolvulaceae, Ebenaceae, Mimosaceae, Caesalpinaceae, Fabaceae, Lagoniaceae, Malvaceae, Palmae, Solanaceae, Tiliaceae, Umbelliferae and Cuscutaceae* (Scherbukhin and Anulov 1999). The genus *Cassia* (*Fabaceae*) is constituted by more than 600 species including bushes, trees and herbs. It is distributed in tropical

Table 7 Some species of genus *Cassia* and their proportion mannose/galactose (Man/Gal)

Species	Man/Gal	Reference
Cassia absus L.	3.0	Kapoor and Mukherjee (1969), (1971), (1972)
Cassia alata L.	3.3	Sen et al. (1987)
Cassia angustifolia Vahl	2.9	Chaubey and Kapoor (2001)
Cassia didymobotrya Fresen.	0.5	Morimoto et al. (1962)
Cassia fastuosa Willd. ex Vogel	4.0	Tavares (1994), Mercê et al. (1998)
Cassia fistula L.	3.0	Kapoor and Farooqui (1993), Morimoto and Unrau (1962)
Cassia grandis L. f.	0.7	Bose and Srivastava (1978), Buckeridge et al. (1995)
Cassia javanica L.	3.23	Azero and Andrade (2002)
Cassia leptocarpa Benth	3.0	Anderson (1949)
Cassia marginata Sessé & Moc.	2.7	Tookey et al. (1962)
Cassia nodosa Buch.-Ham. ex Roxb.	3.5	Heidelberger (1955), Rizvi et al. (1971), Kapoor et al. (1994)
Cassia occidentalis L.	3.0	Gupta and Mukherjee (1973, 1975)
Cassia pulcherrima Dehnh.	3.0	Morimoto and Unrau (1962), Unrau and Choy (1970)
Cassia saltiana Steud.	3.2	Morimoto et al. (1962)
Cassia siamea Lam.	2.5	Kapoor et al. (1996)
Cassia sophera L.	1.0	Hakomori (1964)
Cassia spectabilis DC	2.65	Kapoor et al. (1998)
Cassia tora L.	3.0	Srivastava and Kapoor (2005)
Cassia verrucosa Vogel	4.0	Gupta and Gupta (1988)

regions and subtropical regions around the world (Agarkar and Jadge 1999). Commercial exploitation sources of galactomannans have been identified in some species, whose seeds are of medium size and contain more than 40% endosperm (Srivastava and Kapoor 2005).

Different species differ in the proportion of D-mannose and D-galactose, as well as in the levels of galactomannans in their seeds. The ratio of mannose/galactose is one of the main biochemical characteristics of galactomannans; variations in the monomers provide different physico-chemical properties to their structure (Soni and Bose1985; Buckeridge et al. 1995; Ganter and Reicher 1999).

Galactomannans gums differ from each other according to their different Man/Gal ratios (Table 7). The different chemical properties of these gums make them versatile materials used for many different applications, such as their use as natural coagulants.

8 Limitations, Challenges and Perspective in the Use the Plant-Based Natural Coagulants

Over the years, the minimizing environmental impacts and environmental concern are becoming increasingly important, leading to changes in the energy, water and waste management systems by the development of cleaner production and more efficient, reducing of raw-material consumption and pollution prevention (Nouri et al. 2012; Wu et al. 2010). Plant-based natural coagulants show potential as alternative the usage of chemical coagulants in textile wastewater treatment (Choy et al. 2015). Many studies were mentioned in this Chapter highlighting efficiency plant-based coagulants for coagulation/flocculation process, however, all have some limitations and need to be further studied to identification of fine-tuning and process improvements.

It has been espoused in previous sections that usage of plant-based natural coagulants provides environmental benefits and numerous laboratory-scale have proven that they are technically feasible for small-scale utilization (Yin 2010). But the applications have not been able to transit from laboratory scale to field trials and real industrial applications (Oladoja 2015). CF effectiveness and cost depend on coagulant type and dosage, pH, temperature, ionic strength, total dissolved solids, nature and concentration of organic matter in wastewater, as well as both size and distribution of colloidal particles in suspension and several other factors (Rodrigues et al. 2008; Santo et al. 2012). Therefore, optimization of physical-chemical parameters before large-scale implementation is very important. It takes considerable time and work due to physical-chemical characteristics of real textile wastewater may be subject to periodical variation.

The plant growing and planting at large-scale to supply industrial needs is the main challenge for implementing plant-based natural coagulant at the sustainable industry. This fact will provide an immediate increased in industrial costs due to any necessary technical adaptation, the development of new tools and machines and new professionals needs. But in the long term there will be a significate reduction in the process costs, in the pollutants generated by the process as well it can contribute to employment creation provided by cultivation of those plants.

Another related problem with large-scale is in terms of commercialization. The bottom line is that it will always be based primarily on whether the scale-up system can sustain similar treatment performance at comparable (or reduced) cost with the natural coagulants when compared with established chemical coagulants. The direct comparisons in terms of coagulant types, processing stages and prices in different geographical regions are a very complicated task given the different exchange rates, inflation factor and varying accuracies of the costing values (Yin 2010).

In addition, the response and needs of the market will affect the outlook and demands of natural coagulants as a probable replacement of chemical coagulants. While the commercialization of natural coagulants will not be an overnight success, these small steps taken would progressively bridge the gaps and limitations encountered. Figure 12 summarizes these factors which must be taken into considerations for successful commercialization of natural coagulants. Despite the problems faced in commercialization, many natural coagulants have been used in the water and wastewater treatment industries, especially the tannin extracts from the barks of *Acacia mearnsii* tree have been successfully commercialized as Floccotan®, Tanfloc®, Ecotan® and others (Choy et al. 2014).

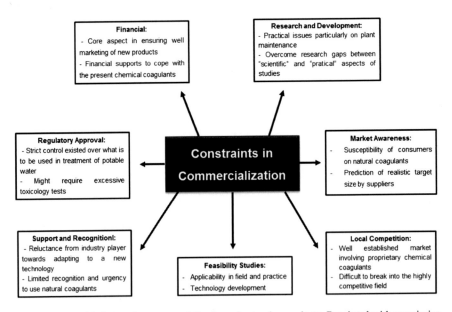

Fig. 12 Factors hindering the commercialization of natural coagulants. Reprinted with permission of Choy et al. (2014). Copyright © 2014 Published by Elsevier B.V.

An evaluation on the economic, social and environmental aspects can be used as a sustainability indicator can be useful with the aid of a pilot plant study. Approval from the local governing bodies should be easily granted for successful commercialization of any new products but they hamper an approval with documentations (Sutherland et al. 2001). Tax rebate, subsidy schemes, tax reduction, incentives environmental laws and regulation should be created by competent bodies to encourage to maximum use of plant-based natural coagulants to resolve the problem.

The dissemination of research findings to the appropriate governmental and non-governmental agencies, who served as an interface between the end users and the researchers, is also lacking. Improvement on this channel of information dissemination can be of great assistance in the implementation of the natural coagulants for real life applications and it also can contribute to resolving the identified problems above (Oladoja 2015).

Understanding the nature of wastewater is fundamental to design suitable wastewater treatment process, to adopt a suitable procedure, determination of acceptable criteria for the residues, determination of a degree of evaluation required to validate the procedure and decision on the residues to be tested based on toxicity (Sahu and Chaudhari 2013). It is required to conduct more and more future research to come up with best natural coagulant or combinations of chemical and natural coagulants. Effectiveness of natural coagulants also required to be carried out against simulated as well as real industrial textile wastewater (Verma et al. 2012). Many studies about simulated wastewater are find in consulted literature but with real wastewater is harder to find.

Despite the sustainable nature of these coagulants, in most cases that the active compound is not isolated and the use of raw extracts has been related, the performance efficiency is low, thus requiring a high dosage of them. Consequent upon the required high optimum dosage, the treated wastewater may contain high levels of residual organic matter; measuring in terms of total organic carbon (TOC) and COD (Oladoja 2016). The substantial increase in the organic load of the treated wastewater is undesirable. It may result to growth in the microbial activities leading the appearance of color, unpleasant odor and producing disagreeable taste (Oladoja 2015).

Conclusive studies about CF mechanism involving natural coagulants are limited and need more research or studies to be understood clearly unlike the widely developed chemical coagulants. The aggregation of colloidal particles may be a result of one or more mechanisms. The procedures for determination and isolation of the active coagulating compounds are too difficult owing to complexity of the process and probable synergistic effects among the components present. Often at times, the active agents responsible for coagulations are made up of different chemical constituents (Choy et al. 2014; Oladoja 2015).

In order to ameliorate the challenges mentioned above, the development of the studies about the CF mechanism can solve them. Thus understanding of the mechanisms of natural coagulants in the CF improves the performance this technique. The characteristics and properties of the flocs formed could reveal important evidence to CF possible mechanisms (Choy et al. 2015). Li et al. (2006) showed that fundamentally different characteristics of flocs such as strength, structure and compactness are essential in establishing the CF mechanisms and the determination of the floc to resistant to breakage is needed to application in industrial scale.

Another factor to be considered to achieve better results in terms CF using natural coagulants is the research of active compound present into plant-based natural coagulant. The extraction, purification, isolation and characterization of the active compound are extremely important to identify the compounds leading to CF. Direct extraction of these compounds could enhance the effectiveness of the overall CF owing to increased relative concentration of the agents. In fact, the coagulant dosage required to achieve optimum coagulation was also significantly reduced by 5–10 times when Marobhe et al. (2007) purified the protein of *V. unguiculata*. This purification also resolves the any problem relating to the presence of the inorganic and organic non-active compounds, which generate an organic load increases (Choy et al. 2014, 2015). For application at large-scale, the plant-based coagulants must be selected according to annual availability of species.

9 Conclusion

The knowledge and scientific research in pollutants removal of wastewater are of utmost importance in order to respond to environmental needs. To meet the increased more and more stringent environmental laws, many different technologies of wastewater treatment have been developed for removal of suspended and dissolved organic matter from textile wastewater. It is evident from a review of current literature that coagulation/flocculation process has been widely used to remove organic matter and color from textile wastewater. The usage of natural coagulants derived from plant-based sources as an alternative to usage of chemical coagulants represents a great development in sustainable environmental technology since it focuses on the improvement of quality of environmental and human health without losing industrial process efficiency. Nonetheless, there are issues that hamper process development of them, such as absence of mass plantation of the plants that affords bulk processing, perceived low-volume market, inexistence of incentive laws and the application is currently restricted to small-scale. In technical terms, these natural coagulants are highly efficient for reduction of physical-chemical parameters of the wastewater such as color, turbidity, COD, TOC, BOD, TSS and others. Planted-based coagulants also have several reasons to become effective coagulant; high cationic charge density, long polymer chains, bridging of aggregates and precipitation, safe, eco-friendly, non-toxic, non-corrosive, high biodegradability, reducing sludge volume, increasing floc size, allowing the reduction of production costs since using raw material from renewable resources.

References

Abiri F, Fallah N, Bonakdarpour B (2016) Sequential anaerobic-aerobic biological treatment of colored wastewaters: case study of a textile dyeing factory wastewater. Water Sci Technol. doi:10.2166/wst.2016.531

Aboulhassan MA, Souabi S, Yaacoubi A, Baudu M (2005) Treatment of textile wastewater using a natural flocculant. Environ Technol 26:705–711

Agarkar SV, Jadge DR (1999) Phytochemical and pharmacological investigations of genus cassia: a review. Asian J Chem 11:295–299

Aguilar MI, Lloréns JSM, Soler A et al (2005) Improvement of coagulation-flocculation process using anionic polyacrylamide as coagulant AID. Chemosphere 58:47–56

Ahmadi N, Chaibakhsh N, Zanjanchi MA (2016) Use of *Descurainia sophia* L. as a natural coagulant for the treatment of dye-containing wastewater. Environ Prog Sustain Energy 35:996–1001. doi:10.1002/ep.12311

Aksu Z, Donmez G (2005) Combined effects of molasses sucrose and reactive dye on the growth and dye bioaccumulation properties of *Candida tropicallis*. Process Bioch 40:2443–2454

Alguacil J, Kauppinen T, Porta N et al (2000) Risk of pancreatic cancer and occupational exposures in Spain. Ann Occup Hyg 44:391–403

Ali H (2010) Biodegradation of synthetic dyes—a review. Water Air Soil Poll 213:251–273. doi:10.1007/s11270-010-0382-4

Anastasakis K, Kalderis D, Diamadopoulos E (2009) Flocculation behavior of mallow and okra mucilage in treating wastewater. Desalination 249:786–791

Anderson E (1949) Endosperm Mucilages of Legumes Ind Eng Chem 41:2887–2890. doi:10.1021/ie50480a056

Annuar MSM, Adnan S, Vikineswary S et al (2009) Kinetics and energetics of azo dye decolorization by *Pycnoporus sanguineus*. Water Air Soil Pollut 202:179–188. doi:10.1007/s11270-008-9968-5

Arslan S, Eyvaz M, Gürbulak E et al (2016) A review of state-of-the-art technologies in dye-containing wastewater treatment—the textile industry case, textile wastewater treatment. Kumbasar EA (ed) InTech. doi:10.5772/64140

Asrafuzzaman M, Fakhruddin ANM, Hossain MA (2011) Reduction of turbidity of water using locally available natural coagulants. ISRN Microbiol 2011:1–6. doi:10.5402/2011/632189

Azero EG, Andrade CT (2002) Testing procedures for galactomannan purification. Polym Test 21:551–556

Bae JS, Freeman HS (2007) Aquatic toxicity evaluation of new direct dyes to the *Daphnia magna*. Dyes Pigments 73:81–85. doi:10.1016/j.dyepig.2005.10.015

Bafana A, Devi SS, Chakrabarti T (2011) Azo dyes: past, present and the future. Environ Rev 19:350–371. doi:10.1139/A11-018

Banks WA, Niehoff ML, Grago D et al (2006) Aluminum complexing enhances amyloid β protein penetration of blood–brain barrier. Brain Res 1116:215–221

Beltrán-Heredia J, Martín JS (2008) Azo dye removal by *Moringa oleifera* seed extract coagulation. Color Technol 124:310–317. doi:10.1111/j.1478-4408.2008.00158.x

Beltrán-Heredia J, Sánchez MJ, Delgado RA et al (2009a) Removal of Alizarin Violet 3R (anthraquinonic dye) from aqueous solutions by natural coagulants. J Hazard Mater 170:43–50. doi:10.1016/j.jhazmat.2009.04.131

Beltrán-Heredia J, Sánchez MJ, Delgado RA (2009b) A removal of carmine indigo dye with *Moringa oleifera* seed extract. Ind Eng Chem Resour 48:6512–6520

Beltrán-Heredia J, Sánchez MJ, Rodríguez SMT (2011) Textile wastewater purification through natural coagulants. Appl Water Sci 1:25–33. doi:10.1007/s13201-011-0005-2

Blanco J, Torrades F, Morón M et al (2014) Photo-Fenton and sequencing batch reactor coupled to photo-Fenton processes for textile wastewater reclamation: feasibility of reuse in dyeing processes. Chem Eng J 240:469–475

Bolto B, Gregory J (2007) Organic polyelectrolytes in water treatment (Revier). Water Res 41:2301–2324

Bose S, Srivastava HC (1978) Structure of a polysaccharide from the seeds of Cassia-grandis lf. 1. Hydrolytic studies. J Indian Chem Soc 16:966–969

Bosetti C, Pira E, LaVecchia C (2005) Bladder cancer risk in painters: a review of the epidemiological evidence, 1989–2004. CCC 16:997–1008

Bratby J (2007) Coagulation and flocculation in water and wastewater treatment, 2nd edn. IWA Publishing

Brookstein DS (2009) Factors associated with textile pattern dermatitis caused by contact allergy to dyes, finishes, foams, and preservatives. Dermatol Clin 27:309–322. doi:10.1016/j.det.2009.05.001

Brüschweiler BJ, Küng S, Bürgi D et al (2014) Identification of non-regulated aromatic amines of toxicological concern which can be cleaved from azo dyes used in clothing textiles. Regul Toxicol Pharm 69:263–272. doi:10.1016/j.yrtph.2014.04.011

Buckeridge MS, Panegassi VR, Rocha DC et al (1995) Seed galactomannan and evolution of the Leguminosae. Phytochemistry 38:871–875

Castanho M, Malpass GRP, Motheo AJ (2006) Evaluation of electrochemical and photoelectro-chemical methods for the degradation of three textile dyes. Quim Nova 29:983–989

Castro FD, Bassin JP, Dezotti M (2016) Treatment of a simulated textile wastewater containing the Reactive Orange 16 azo dye by a combination of ozonation and moving-bed biofilm reactor: evaluating the performance, toxicity, and oxidation by-products. Environ Sci Pollut Res 1–100. doi:10.1007/s11356-016-7119-x

Chaibakhsh N, Ahmadi N, Zanjanchi MA (2014) Use of Plantago major L. as a natural coagulant for optimized decolorization of dye-containing wastewater. Ind Crop Prod 61:169–175. doi:10.1016/j.indcrop.2014.06.056

Chaubey M, Kapoor VP (2001) Structure of a galactomannan from the seeds of Cassia angustifolia Vahl. Carbohyd Res 332:439–444

Chequer FMD, Dorta DJ, Oliveira DP (2011) Azo dyes and their metabolites: does the discharge of the azo dye into water bodies represent human and ecological risks? Advances in treating textile effluent. In: Hauser P (ed) InTech, doi:10.5772/19872

Chethana M, Sorokhaibam LG, Bhandari MV et al (2015) Application of biocoagulant Acanthocereus tetragonus (Triangle cactus) in dye wastewater treatment. J Environ Res Dev 9:813

Chethana M, Sorokhaibam LG, Bhandari MV et al (2016) Green approach to dye wastewater treatment using biocoagulants. ACS Suntain Chem Eng 4:2495–2507. doi:10.1021/acssuschemeng.5b01553

Chhabra M, Mishra S, Sreekrishnan TR (2015) Combination of chemical and enzymatic treatment for efficient decolorization/degradation of textile effluent: High operational stability of the continuous process. Biochem Eng J 93:17–24. doi:10.1016/j.bej.2014.09.007

Choy SY, Prasad KMN, WU TY et al (2014) Utilization of plant-based natural coagulants as future alternatives towards sustainable water clarification. J Environ Sci 26:2178–2189. doi:10.1016/j.jes.2014.09.024

Choy SY, Prasad KMN, Wu TY et al (2015) A review on common vegetables and legumes as promising plant-based natural coagulants in water clarification. Int J Environ Sci Technol 12:367–390. doi:10.1007/s13762-013-0446-2

Choy SY, Prasad KMN, Wu TY et al (2016) Performance of conventional starches as natural coagulants for turbidity removal. Ecol Eng 94:352–364. doi:10.1016/j.ecoleng.2016.05.082

Chung KT, Fulk GE, Egan M (1978) Reduction of azo dyes by intestinal anaerobes. Appl Environ Microb 35:558–562

Colpini LMS, Alves HJ, Santos OAA et al (2008) Discoloration and degradation of textile dye aqueous solutions with titanium oxide catalysts obtained by the sol-gel method. Dyes Pigments 76:525–529. doi:10.1016/j.dyepig.2006.10.014

Colpini LMS, Lenzi GG, Urio MB et al (2014) Photodiscoloration of textile reactive dyes on Ni/TiO$_2$ prepared by the impregnation method: effect of calcination temperature. J Environ Chem Eng 2:2365–2371. doi:10.1016/j.jece.2014.01.007

Dasgupta J, Sikder J, Chakraborty S et al (2015) Remediation of textile effluents by membrane based treatment techniques: a state of the art review. J Environ Manage 147:55–72. doi:10.1016/j.jenvman.2014.08.008

De Roos AJ, Gao DL, Wernli KJ et al (2005) Colorectal cancer incidence among female textile workers in Shanghai, China: a case-cohort analysis of occupational exposures. CCC 16:1177–88

De Souza MTF, Ambrosio E, Almeida CA et al (2014) The use of a natural coagulant (*Opuntia ficus-indica*) in the removal for organic materials of textile effluents. Environ Monitor Assess 186:5261–5271. doi:10.1007/s10661-014-3775-9

De Souza MTF, Almeida CA, Ambrosio E et al (2016a) Extraction and use of *Cereus peruvianus* cactus mucilage in the treatment of textile effluents. J Taiwan Inst Chem 67:174–183. doi:10.1016/j.jtice.2016.07.009

De Souza PC, Pereira NC, Gonçalves MS et al (2016b) Study of the treatment of textile effluent by coagulation/flocculation process and electrocoagulation. E-xacta 9:123–132. doi:10.18674/exacta.v9i2.195S

Dellamatrice PM, Monteiro RTR, Balan DSL (2008) Biodegradação de corantes têxteis. Microbiologia Ambiental, 2nd edn., Jaguariúna: Embrapa e Meio Ambiente, pp 321–338

Deveci EÜ, Dizge N, Yatmaz HC et al (2016) Integrated process of fungal membrane bioreactor and photocatalytic membrane reactor for the treatment of industrial textile wastewater. Biochem Eng J 105:420–427. doi:10.1016/j.bej.2015.10.016

Dolin PJ (1992) A descriptive study of occupation and bladder cancer in England & Wales. Br J Cancer 66:568–578

Duan J, Niu A, Shi D et al (2009) Factors affecting the coagulation of seawater by ferric chloride. Desalin Water Treat 11:173–183

Eckenfelder WW (1966) Industrial water pollution control. Mc Graw-Hill, New York

Edilson TNJI, Atchudan R, Sethuraman MG et al (2016) Reductive-degradation of carcinogenic azo dyes using *Anacardium occidentale* testa derived silver nanoparticles. J Photochem Photobiol B 162:604–610

Ferraz ERA, Oliveira GAR, Grando MD et al (2013) Photoelectrocatalysis based on Ti/TiO$_2$ nanotubes removes toxic properties of the azo dyes Disperse Red 1, Disperse Red 13 and Disperse Orange 1 from aqueous chloride samples. J Environ Manage 124:108–114. doi:10.1016/j.jenvman.2013.03.033

Foo KY, Hameed BH (2010) Decontamination of textile wastewater via TiO$_2$/activated carbon composite materials. Adv Colloid Interface Sci 159:130–143. doi:10.1016/j.cis.2010.06.002

Freitas TKFS, Oliveira VM, De SOUZAMTF et al (2015) Optimization of coagulation-flocculation process for treatment of industrial textile wastewater using okra (*A. esculentus*) mucilage as natural coagulant. Ind Crops Prod 76:538–544

Fu F, Wang Q (2011) Removal of heavy metal ions from wastewaters: a review. J Environ Manage 92:407–418. doi:10.1016/j.jcnvman.2010.11.011

Ganter JLMS, Reicher F (1999) Watersoluble galactomannans from seeds of *Mimosaceae* spp. Bioresour Technol 68:55–62

Garcia JC, Simionato JI, da Silva AEC et al (2009) Solar photocatalytic degradation of real textile effluents by associated titanium dioxide and hydrogen peroxide. Sol Energy 83:316–322. doi:10.1016/j.solener.2008.08.004

Garcia JC, Freitas TKF, Palácio SM et al (2012) Toxicity assessment of textile effluents treated by advanced oxidative process (UV/TiO$_2$ and UV/TiO$_2$/H$_2$O$_2$) in the species *Artemia salina* L. Environ Monit Assess 185:2179–2187. doi:10.1007/s10661-012-2698-6

Góes MM, Keller M, Oliveira VM et al (2016) Polyurethane foams synthesized from cellulose-based wastes: Kinetics studies of dye adsorption. Ind Crops Prod 85:149–158

Gottlieb A, Shaw C, Smith A et al (2003) The toxicity of textile reactive azo dyes after hydrolysis and decolourisation. J Biotechnol 101:49–56

Gozálvez-Zafrilla JM, Sanz-Escribano D, Lora-García D et al (2008) Nanofiltration of secondary effluent for wastewater reuse in the textile industry. Desalination 222:272–279. doi:10.1016/j. desal.2007.01.173

Guaratini CCI, Zanoni MVB (2000) Textile Dyes Quim Nova 23:71–78

Gümüş D, Akbal F (2011) Photocatalytic degradation of textile dye and wastewater. Water Air Soil Pollut 216:117–124. doi:10.1007/s11270-010-0520-z

Gupta R, Gupta PC (1988) A neutral seed-gum from *Crotalaria verrucosa*. Carbohyd Res 181:287–292

Gupta DS, Mukherjee S (1973) Structure of a galactomannan from *Cassia occidentalis* seed. India J Chem 11:1134–1137

Gupta DS, Mukherjee S (1975) Structure of galactomannan from *Cassia occidentalis* seed: isolation & structure elucidation of oligosaccharides. India J Chem 13:1152–1154

Gupta VK, Khamparia S, Tyagi I et al (2015) Decolorization of mixture of dyes: a critical review. Global J Environ Sci Manage 1:71–94. doi:10.7508/gjesm.2015.01.007

Hakomori S (1964) A rapid permethylation of glycolipid, and polysaccharide catalyzed by methylsulfinyl carbanion in dimethyl sulfoxide. J Biochem 55:205

Hanif MA, Nadeem R, Zafar MN et al (2008) Physical-chemical treatment of textile wastewater using natural coagulant *Cassia fistula* (Golden Shower) Pod Biomass. J Chem Soc Pak 30:385–393

Hassani AH, Seif S, Javid AH et al (2008) Comparison of adsorption process by GAC with novel formulation of coagulation flocculation for color removal of textile wastewater. Int J Environ Res 2:239–248

Heidelberger M (1955) Immunological specificities involving multiple units of galactose. Il1. J Am Chem Soc 77:4308

Hemapriya G, Abinaya R, Dhinesh Kumar S (2015) Textile effluent treatment using *Moringa oleifera*. IJIRD 4:385–390

Holkar CR, Jadhav AJ, Pinjari DV et al (2016) A critical review on textile wastewater treatments: Possible approaches. J Environ Manage 182:351–366. doi:10.1016/j.jenvman.2016.07.090

Holt PK, Barton GW, Wark M et al (2002) A quantitative comparison between chemical dosing and electrocoagulation. Colloids Surf A Physicochem Eng Asp 211:233–248

Hunger K (2007) Industrial dyes: chemistry, properties, Applications. Wiley-VCH, Weinheim, Cambridge

Jegatheesan V, Pramanik BK, Chen J et al (2016) Treatment of textile wastewater with membrane bioreactor: a critical review. Bioresour Technol 204:202–212. doi:10.1016/j.biortech. 2016.01.006

Jekel M (1997) Wastewater treatment in the Textile Industry. In: Treatment of wastewaters from textile processing. TU, Berlin. Schriftenreihe Biologische Abwasscrrciigung des Sfb 193, Berlin

Jeon JR, Kim EJ, Kim JM et al (2009) Use of grape seed and its natural polyphenol extracts as a natural organic coagulant for removal of cationic dyes. Chemosphere 77:1090–1098. doi:10. 1016/j.chemosphere.2009.08.036

Kakoi B, Kaluli JW, Ndiba P et al (2016) *Banana pith* as a natural coagulant for polluted river water. Ecol Eng 95:699–705. doi:10.1016/j.ecoleng.2016.07.001

Kalia S, Averous L (2011) Biopolymers and environmental applications. Wiley, Hoboken, p 642

Kansal S, Kumari A (2014) Potential of *M. oleifera* for the treatment of water and wastewater. Chem Rev 114:4993–5010. doi:10.1021/cr400093w

Kant R (2012) Textile dyeing industry an environmental hazard. Nat Sci 4:22–26. doi:10.4236/ns. 2012.41004

Kapoor VP, Farooqui MIH (1993) Cassia-fistula (amaltas) seed-a potential source of commercial gum. Res Ind 38:3

Kapoor VP, Mukherjee S (1969) Nature of sugars in *Cassia absus* seed polysaccharide. Curr Sci (India) 38:38

Kapoor VP, Mukherjee S (1971) Isolation of five oligosaccharides from the galactomannan of Cassia absus seed. Phytochemistry 10:655

Kapoor VP, Mukherjee S (1972) Galactomannan from *Cassia absus* seed. 3. structure of degraded galactomannan. Indian J Chem 10:155

Kapoor VP, Milas M, Taravel FR et al (1994) Rheological properties of a seed galactomannan form Cassia nodosa buch-hem. Carbohyd Polym 25:79–84

Kapoor VP, Milas M, Taravel FR et al (1996) Rheological properties of a seed galactomannan from *Cassia siamea* Lamk. Food Hydrocoll 10:167–172

Kapoor VP, Taravel FR, Joseleau J et al (1998) *Cassia spectabilis* DC seed alactomannan: structural, crystallographical and rheological studies. Carbohyd Res 306:231–241

Khaled A, El Nemr A, El-Sikaily A et al (2009) Treatment of artificial textile dye effluent containing Direct Yellow 12 by orange peel carbon. Desalination 238:210–232. doi:10.1016/j.desal.2008.02.014

Khan R, Bhawana P, Fulekar MH (2013) Microbial decolorization and degradation of synthetic dyes: a review. Rev Environ Sci Biotechnol 12:75–97. doi:10.1007/s11157-012-9287-6

Khandegar V, Saroha AK (2013) Electrocoagulation for the treatment of textile industry effluent—a review. J Environ Manage 128:949–963. doi:10.1016/j.jenvman.2013.06.043

Kiernan JA (2001) Classification and naming of dyes, stains and fluorochromes. Biotech Histochem 76:261–277. doi:10.1080/bih.76.5-6.261.278

Kunz A, Peralta-Zamora P, Moraes SG et al (2002) New tendencies on textile effluent treatment. Quim Nova 25:78–82

Leme FP (1979) Teoria e técnicas de tratamento de água São Paulo. CETESB, SP

Leme FP (1982) Engenharia do saneamento ambiental. LTC – Livros Técnicos e Científicos, Rio de Janeiro

Leme DM, Oliveira GAR, Meireles G et al (2014) Genotoxicological assessment of two reactive dyes extracted from cotton fibres using artificial sweat. Toxicol In Vitro 28:31–38. doi:10.1016/j.tiv.2013.06.005

Lenzi E, Favero LOB, Luchese EB (2009) Introdução à Química da água: ciência, vida e sobrevivência. LTC, Rio de Janeiro

Li T, Zhu Z, Wang D et al (2006) Characterization of f loc size, strength and structure under various coagulation mechanisms. Powder Technol 168:104–110

Lozano-Rivas WA, Whiting KE, Gómez-Lahoz C 92016) Use of glycosides extracted from the fique (*Furcraea* sp.) in wastewater treatment for textile industry. Int J Environ Sci Technol 13:1131–1136

Manh HB, Giang HDT, Ngoc TL (2015) Decolorization of Reactive Red 195 solution by Cassia fistula seed gum. Sci Technol Devel 18:5–11

Marobhe NJ, Dalhammar G, Gunaratna KR (2007) Simple and rapid methods for purification and characterization of active coagulants from the seeds of Vigna unguiculata and Parkinsonia aculeata. Environ Technol 28(6):671–681

Mastrangelo G, Fedeli U, Fadda E et al (2002) Epidemiologic evidence of cancer risk in textile industry workers: a review and update. Toxicol Ind Health 18:171–181

Mathur N, Bhatnagar P, Nagar P et al (2005) Mutagenicity assessment of effluents from textile/dye industries of Sanganer, Jaipur (India): a case study. Ecotox Environ Safe 61:105–113. doi:10.1016/j.ecoenv.2004.08.003

Matilainen A, Vepsäläinen M, Sillanpää M (2010) Natural organic matter removal by coagulation during drinking water treatment: a review. Adv Colloid Interface Sci 159:189–197. doi:10.1016/j.cis.2010.06.007

McLachlan CRD (1995) Aluminium and the risk for Alzheimer's Disease. Environmetrics 6:233–275

Mercê ALR, Lombardi SC, Mangrich AS et al (1998) Equilibrium studies of galactomannan of Cassia fastuosa and Leucaena leucocephala and Cu^{2+} using potentiometry and EPR spectroscopy. Carbohyd Polym 35:13–20

Merzouk B, Madani K, Sekki A (2010) Using electrocoagulation–electroflotation technology to treat synthetic solution and textile wastewater, two case studies. Desalination 250:573–577. doi:10.1016/j.desal.2009.09.026

Miller SM, Fugate EJ, Craver VO et al (2008) Toward understanding the efficacy and mechanism of Opuntia spp. as a natural coagulant for potential application in water treatment. Environ. Sci Technol 42:4274–4279

Mishra A, Bajpai M (2005) Flocculation behaviour of model textile wastewater treated with a food grade polysaccharide. J Hazard Mater B118:213–217. doi:10.1016/j.jhazmat.2004.11.003

Mishra A, Bajpai M (2006) The flocculation performance of Tamarindus mucilage in relation to removal of vat and direct dyes. Bioresour Technol 97:1055–1059. doi:10.1016/j.biortech.2005. 04.049

Mishra A, Srinivasan R, Dubey R (2002) Flocculation of textile wastewater by plantago psyllium Mucilage. Macromol Mater Eng 287:592–596

Moraes SG, Freire RS, Durán N (2000) Degradation and toxicity reduction of textile effluent by combined photocatalytic and ozonation processes. Chemosphere 40:369–373. doi:10.1016/ S0045-6535(99)00239-8

Morimoto JY, Unrau AM (1962) Observations on the gums (galacto-raannans) of some legume seeds. Hawaii Farm Sci 11:6

Morimoto TY, Unrau ICJ, Urnau AMJ (1962) Legume polysaccharides, chemical and physical properties and the enzymatic degradation of some tropical plant gums. J Agric Food Chem 10:134

Moura JCVP (2003) Síntese de corantes e aplicação a fibras têxteis. Química 75–79

Naje AS, Chelliapan S, Zakaria Z et al (2016) Electrocoagulation using a rotated anode: A novel reactor design for textile wastewater treatment. J Environ Manage 176:34–44. doi:10.1016/j. jenvman.2016.03.034

Ndabigengesere A, Subba Narasiah K, Talbot BG (1995) Active agents and mechanism of coagulation of turbid waters using Moringa oleifera. Water Res 29:703–710. doi:10.1016/ 0043-1354(94)00161-Y

Notani PN, Shah P, Balajrishnan V (1993) Occupation and cancers of the lung and bladder: a case-control study in Bombay. Int J Epidemiol 22:185–191

Nouri J, Nouri N, Moeeni M (2012) Development of industrial waste disposal scenarios using life-cycle assessment approach. Int J Environ Sci Technol 9(3):417–424

O'Neill C, Hawkes FR, Hawkes DL et al (1999) Colour in textile effluents – sources, measurement, discharge consents and simulation: a review. J Chem Technol Biot 74:1009–1018. doi:10.1002/(SICI)1097-4660(199911)74:11<1009:AID-JCTB153>3.0.CO;2-N

Oladoja NA (2015) Headway on natural polymeric coagulants in water and wastewater treatment operations. J Water Process Eng 6:174–192. doi:10.1016/j.jwpe.2015.04.004

Oladoja NA (2016) Advances in the quest for substitute for synthetic organic polyelectrolytes as coagulant aid in water and wastewater treatment operations. Sustain Chem Pharm 3:47–58. doi:10.1016/j.scp.2016.04.001

Oliveira GAR, Lapuente J, Teixidó E et al (2016) Textile dyes induce toxicity on Zebrafish early life stages. Environ Toxicol Chem 35:429–434. doi:10.1002/etc.3202

Ozyonar F (2015) Optimization of operational parameters of electrocoagulation process for real textile wastewater treatment using Taguchi experimental design method. Desalin Water Treat 1–11. doi:10.1080/19443994.2015.1005153

Palácio SM, Espinoza-Quiñones FR, Módenes AN et al (2012) Optimised photocatalytic degradation of a mixture of azo dyes using a $TiO_2/H_2O_2/UV$ process. Water Sci Technol 65:1392–1398. doi:10.2166/wst.2012.015

Patel H, Vashi RT (2012) Removal of Congo Red dye from its aqueous solution using natural coagulants. J Saudi Chem Soc 16:131–136. doi:10.1016/j.jscs.2010.12.003

Peralta-Zamora P, Tiburtius ERL, Moraes SG et al (2002) Degradação Enzimática de Corantes Têxteis. Revista Química Têxtil 23:32–38

Perng YS, Bui MH (2015) The feasibility of Cassia fistula gum with polyaluminum chloride for the decolorization of reactive dyeing wastewater. J Serb Chem Soc 80:115–125

Platzek T (2010) Risk from exposure to arylamines from consumer products and hair dyes. Front Biosci E2(1):1169–1183. doi:10.2741/E177

Polizzi S, Pira E, Ferrara M et al (2002) Neurotoxic effects of aluminium among foundry workers and Alzheimer's disease. NeuroToxicol 23:761–774

Prasad SVM, Rao BS (2016) Influence of plant-based coagulants in waste water treatment. IJLTEMAS 5:45–48

Ramavandi B, Farjadfard S (2014) Removal of chemical oxygen demand from textile wastewater using a natural coagulant. Korean J Chem Eng 31:81–87. doi:10.1007/s11814-013-0197-2

Rasteiro MG, Garcia FAP, Ferreira P et al (2008) The use of LDS as a tool to evaluate flocculation mechanisms. Chem Eng Prog 47:1323–1332

Reid GJS (1985) Galactomannans. Biochemistry of storage carbohydrates. Academic Press, London, pp 265–286

Reid GJS, Edwards ME (1995) Galactomannans and other cell wall storage polysaccharides in seeds. Food polysaccharides and their applications. New York, pp 155–186

Renault F, Sancey B, Charles J et al (2009) Chitosan flocculation of cardboard-mill secondary biological wastewater. Chem Eng J 155:775–783

Rizvi SAI, Gupta PC, Kaul RK (1971) The structure of a galactomannan from the seeds of Cassia nodosa. Planta Med 20:24–32

Robinson T, McMullan G, Marchant R et al (2001) Remediation of dyes in textile effluent: a critical review on current treatment technologies with a proposed alternative. Bioresour Technol 77:247–255

Rodrigues AC, Boroski M, Shimada NS et al (2008) Treatment of paper pulp and paper Mill wastewater by coagulation-floculation followed by heterogeneous photocatalysis. J Photochem Photobiol A Chem 194:1–10. doi:10.1016/j.jphotochem.2007.07.007

Ruehrwein RA, Ward DW (1952) Mechanism of clay aggregation by polyelectrolytes. Soil Sci 73:485–492

Sahu OP, Chaudhari PK (2013) Review on Chemical treatment of Industrial Waste Water. J Appl Sci Environ Manage 17:241–257

Sánchez-Gilo A, Gómez-De La Fuente E, Calzado L et al (2010) Textile contact dermatitis in a patient sensitized to reactive orange 107 Dye. Actas Dermosifiliogr 101:278–279

Sanghi R, Verma P (2013) Decolorisation of aqueous dye solutions by low-cost adsorbents: a review. Color Technol 129:85–108

Sanghi R, Bhatttacharya B, Singh V (2002) Cassia angustifolia seed gum as an effective natural coagulant for decolourisation of dye solutions. Green Chem 4:252–254. doi:10.1039/b200067a

Sanghi R, Bhattacharyaa B, Dixita A et al (2006a) Ipomoea dasysperma seed gum: an effective natural coagulant for the decolorization of textile dye solutions. J Environ Manage 81:36–41. doi:10.1016/j.jenvman.2005.09.015

Sanghi R, Bhatttacharya B, Singh V (2006b) Use of Cassia javahikai seed gum and gum-g-polyacrylamide as coagulant aid for the decolorization of textile dye solutions. Bioresour Technol 97:1259–1264. doi:10.1016/j.biortech.2005.05.004

Santo CE, Vilar VJP, Botelho MS et al (2012) Optimization of coagulation-flocculation and flotation parameters for the treatment of a petroleum refinery effluent from a Portuguese plant. Chem Eng J 188:117–123. doi:10.1016/j.cej.2011.12.041

Santos AB, Cervantes FJ, Van Lier JB (2007) Review paper on current technologies for decolourisation of textile wastewaters: perspectives for anaerobic biotechnology. Bioresour Technol 98:2369–2385. doi:10.1016/j.biortech.2006.11.013

Santos LB, Domingues FS, Rosseto F et al (2013) Simultaneous determination of textile dyes by adsorptive cathodic stripping voltammetry. Acta Sci-Technol 35:387–392. doi:10.4025/actascitechnol.v35i2.16595

Santos LB, Paulino AT, Domingues FS et al (2016) Derivative cathodic striping voltammetry in the simultaneous determination of three textile dyes in aqueous solutions. Color Technol 132:201–207. doi:10.1111/cote.12209

Saratale RG, Saratale GD, Chang JS et al (2011) Bacterial decolorization and degradation of azo dyes: a review. J Taiwan Inst Chem Eng 42:138–157. doi:10.1016/j.jtice.2010.06.006

Scherbukhin VD, Anulov OV (1999) Legume seed galactomannans (review). Appl Biochem Microbiol 35:229–244

Sen AK, Sarkar KK, Banerjee N et al (1987) Structural investigation of a water-soluble galactomannan from the seeds of *Cassia-alata linn Farooqui*. Indian J Chem, Sect B 26:2–25

Sen SK, Raut S, Bandyopadhyay P et al (2016) Fungal decolouration and degradation of azo dyes: A review. Fungal Biol Rev 30:112–133

Senthilkumar M, Gnanapragasam G, Arutchelvan V et al (2011) Treatment of textile dyeing wastewater using two-phase pilot plant UASB with sago wastewater as co-substrate. Chem Eng J166:10–14. doi:10.1016/j.cej.2010.07.057

Shamsnejati S, Chaibakhshb N, Pendashtehc AR et al (2015) Mucilaginous seed of *Ocimum basilicum* as a natural coagulant for textile wastewater treatment. Ind Crop Prod 69:40–47

Sharma BR, Dhuldhoya NC, Merchant UC (2006) Flocculants—an ecofriendly approach. J Polym Environ 14:195–202. doi:10.1007/s10924-006-0011-x

Sher F, Malik A, Liu H (2013) Industrial polymer effluent treatment by chemical coagulation and flocculation. J Environ Chem Eng 1:684–689

Simionato JI, Villalobos LDG, Bulla MK et al (2014) Application of chitin and chitosan extracted from silkworm chrysalides in the treatment of textile effluents contaminated with remazol dyes. Acta Sci-Technol 36:693–698. doi:10.4025/actascitechnol.v36i4.24428

Soares PA, Souza R, Soler J et al (2017) Remediation of a synthetic textile wastewater from polyester-cotton dyeing combining biological and photochemical oxidation processes. Sep Purif Technol 172:450–462. doi:10.1016/j.seppur.2016.08.036

Soni SK, Bose S (1985) Seed galactomannans & Their structures. J Sci Ind Res 44:544–547

Souza RP, Freitas TKFS, Domingues FS et al (2016) Photocatalytic activity of TiO_2, ZnO and Nb_2O_5 applied to degradation of textile wastewater. J Photochem Photobiol A Chem 329:9–17. doi:10.1016/j.jphotochem.2016.06.013

Souza RP, Ambrosio E, Souza MTF et al (2017) Solar photocatalytic degradation of textile effluent with TiO2, ZnO, and Nb2O5 catalysts: assessment of photocatalytic activity and mineralization. Environ Sci Pollut Res 1–9. doi:10.1007/s11356-017-8408-8

Sponza DT (2006) Toxicity studies in a chemical dye production industry in Turkey. J Hazard Mater 138:438–447. doi:10.1016/j.jhazmat.2006.05.120

Srivastava M, Kapoor VP (2005) Seed galactomannans: an overview. Chem Biodivers 2:295–317

Subramonian W, Wu TY, Chai S (2014) A comprehensive study on coagulant performance and floc characterization of natural *Cassia obtusifolia* seed gum in treatment of raw pulp and paper mill effluent. Ind Crop Prod 61:317–324

Sutherland J, Folklard G, Poirier Y (2001) Moringa oleifera. The constraints to commercialization. International Workshop: 29th October–2nd November, Dar es Salaam, Tanzania

Tavares GA (1994) Estrutura e propriedades físico-químicas da galactomanana de sementes de *Cassia fastuosa Willd. (Cassia)*. Dissertation, Federal University of Paraná

Teh CY, Budiman PM, Shak KPY et al (2016) Recent advancement of coagulation-flocculation and its application in wastewater treatment. Ind Eng Chem Res 55:4363–4389

Tie J, Jiang M, Li H et al (2015) A comparison between *Moringa oleifera* seed presscake extract and polyaluminum chloride in the removal of direct black 19 from synthetic wastewater. Ind Crop Prod 74:530–534. doi:10.1016/j.indcrop.2015.04.004

Tookey HL, Lohmar RL, Wolf IA et al (1962) Seed polysaccharides, new sources of seed mucilages. Agric Food Chem 10:131–133

Uday USP, Bandyopadhyay TK, Bhunia B (2016) Bioremediation and detoxification technology for treatment of dye(s) from textile effluent, textile wastewater treatment. In: Kumbasar EA (ed) InTech. doi:10.5772/62309

Unrau AM, Choy YM (1970) Identification of linkages of a galactomannan isolated from seed of *Caesalpinia pulcherima*. Carbohyd Res 14:151–158

Valh JV, Le Marechal AM, Vajnhandl S et al (2011) Water in the Textile Industry. Treatise Water Sci Elsevier 4:685–706. doi:10.1016/b978-0-444-53199-5.00102-0

Veeramalini JB, Sravanakumar K, Joshua Amarnath D (2012) Removal of reactive yellow dye from aqueous solutions by using natural coagulant *Moringa oleifera*. IJSET 1:56–62

Verma AK, Dash RR, Bhunia P (2012) A review on chemical coagulation/flocculation technologies for removal of colour from textile wastewaters. J Environ Manage 93:154–168. doi:10.1016/j.jenvman.2011.09.012

Vijayaraghavan G, Shanthakumar S (2015) Removal of sulphur black dye from its aqueous solution using alginate from *Sargassum* sp. (brown algae) as a coagulant. Environ Prog Sustain Energy 34:1427–1434. doi:10.1002/ep.12144

Vineis P, Pirastu R (1997) Aromatic amines and cancer. CCC 8:346–355

Vishali S, Karthikeyan R (2015) *Cactus opuntia* (ficus-indica): an eco-friendly alternative coagulant in the treatment of paint effluent. Desalin Water Treat 56:1489–1497

Vishali S, Rashmi P, Karthikeyan R (2016) Evaluation of wasted biomaterial, crab shells (Portunus sanguinolentus), as a coagulant, in paint effluent treatment. Desalin Water Treat 57:13157–13165

Walton JR (2013) Aluminum's involvement in the progression of Alzheimer's disease. J Alzheimers Di 35:7–43. doi:10.3233/JAD-121909

Weber E, Wolfe NL (1987) Kinetics studies of reduction of aromatic azo compounds in anaerobic sediment/water systems. Environ Toxicol Chem 6:911–920

Wernli KJ, Fitzgibbons ED, Ray RM et al (2006) Occupational risk factors for esophageal and stomach cancers among female textile workers in Shanghai, China. Am J Epidemiol 163:717–25

WHO—World Health Organization (2003) Acrylamide in drinking water. http://www.who.int/water_sanitation_health/dwq/chemicals/acrylamide.pdf. Accessed 20 Dec 2016

Wu TY, Mohammad AW, Jahim JM, Anuar N (2010) Pollution control technologies for the treatment of palm oil mill effluent (POME) through end-of-pipe processes. J Environ Manag 91 (7):1467–1490

Xu XR, Li HB, Wang WH et al (2005) Decolorization of dyes and textile wastewater by potassium permanganate. Chemosphere 59:893–898. doi:10.1016/j.chemosphere.2004.11.013

Yagub MT, Sen TK, Afroze S et al (2014) Dye and its removal from aqueous solution by adsorption: a review. Adv Colloid Interface Sci 209:172–184. doi:10.1016/j.cis.2014.04.002

Yin CY (2010) Emerging usage of plant-based coagulants for water and wastewater treatment. Process Biochem 45:1437–1444. doi:10.1016/j.procbio.2010.05.030

Zahrim AY, Tizaoui C, Hilal A (2011) Coagulation with polymers for nanofiltration pre-treatment of highly concentrated dyes: a review. Desalination 266:1–16. doi:10.1016/j.desal.2010.08.012

Zhang J, Zhang F, Luo Y et al (2006) A preliminary study on cactus coagulant in water treatment. Process Biochem 41:730–733. doi:10.1016/j.procbio.2005.08.016

Zhao S, Gao B, Yue Q et al (2014) Study of Enteromorpha polysaccharides as a new-style coagulant aid in dye wastewater treatment. Carbohyd Polym 103:179–186. doi:10.1016/j.carbpol.2013.12.045

Zhu G, Zheng H, Zhang Z et al (2011) Characterization and coagulation–flocculation behavior of polymeric aluminum ferric sulfate (PAFS). Chem Eng J 178:50–59. doi:10.1016/j.cej.2011.10.008

Zollinger H (1991) Color chemistry: syntheses, properties and applications of organic dyes and pigments, 2nd edn. VCH, New York

New Textile Waste Management Through Collaborative Business Models for Sustainable Innovation

Armaghan Chizaryfard, Yasaman Samie and Rudrajeet Pal

Abstract In most nations, textile waste management is recognized to be a multi-actor system; however most participating actors tend to play a significant role in handling and treating the textile waste single-handedly thus resulting in a very fragmented system fraught with many challenges. In addition, the main textile waste treatment, e.g. in Sweden is still incineration (nearly 55% of per capita disposal) resulting in low degrees of value generation. Nearly 20% of the waste is handled by ten major charities in Sweden. This highlights the necessity for the actors to perform in a network and expand their collaboration, thus move more efficiently towards development of a sustainable value innovation, and find an alternative new way to manage textile waste. Given this our study strives to investigate the challenges and opportunities of implementation of a collaborative business model for sustainable innovation. By taking the benefits of actor-, activity- and value-mapping technique, our study helps in gaining a better realization of the Swedish textile waste management system. The core values of actors have been identified along with the identification of their shared and conflicting values with the aid of a value mapping tool. Data was collected through semi-structured interviews from seven organizations representing the Swedish textile waste management system. Overall our study provides a rich and descriptive picture of the participating actors, their activities, collaboration and value-orientations within the Swedish textile waste management system, and highlights the key drivers of a collaborative solution, viz. legislation, trust and shared understanding and communication, that can be foreseen to increase dialogue and collaboration among actors to support the movement from egocentric to a multi-actor business model. A clear benefit of such collaborative business models is substitution of incineration by higher degrees of reuse of textiles, which has high potential to generate positive environmental impact, through reduction of toxic effects of textile incineration and also new production processes.

A. Chizaryfard · Y. Samie · R. Pal (✉)
Department of Business Administration and Textile Management,
The Swedish School of Textiles, University of Borås, Borås, Sweden
e-mail: rudrajeet.pal@hb.se

© Springer Nature Singapore Pte Ltd. 2018
S.S. Muthu (ed.), *Detox Fashion*, Textile Science and Clothing Technology,
DOI 10.1007/978-981-10-4780-0_3

81

Keywords Waste management · Textile waste · Value mapping · Collaborative business model · Sustainable innovation · Sweden

1 Introduction

Increase in consumption of textiles has introduced environmental challenges in regards to production, use and finally end-of-life management of textile materials during the last decades (Watson et al. 2016). In the case of Sweden, the consumption has grown 40% over the last ten years which results in greater amount of textile waste (Naturvårdsverket 2013). This highlights the need for a well-developed and efficient textile waste management (TWM) system. Incineration is currently the predominant activity for treatment of textile waste in Sweden however it is favourable to replace it by alternative activities higher up along the EU waste hierarchy, due to the high global warming potential and primary energy usage inherent to incineration (Zamani 2012). This calls for deeper investigation leading to the development of innovative solutions for TWM. However, recent studies (e.g. Palm 2011) show that economic hindrance act as a major barrier towards operating a more sustainable TWM system. In the current value chain structure, product costing is debarred of environmental cost considerations (which would be sufficiently higher than what it is) if the finite and limited availability of virgin textiles and raw material resources is taken into account. However, a lack of such considerations leads to market prices of reused and recycled textiles to be considerably high, thus resulting in a lack of economic viability of the current TWM system. Further, the production of virgin textiles is run in low cost countries while collection of the used textiles is operated in Sweden where the labour cost is remarkably high. This is one of the major reasons why there is a lack of large scale and consolidated efforts towards TWM in high-cost nations, such as Sweden. Palm (2011) along this line suggests that an improved TWM requires new cost efficient methods for textile recycling to enable high grade of recycling of textiles, aimed primarily to fetch high price points in the market. Moreover, other factors contributing to this challenge are related to inefficient and fragmented structure of the TWM system in Sweden (for collection and sorting) and a lack of advanced recycling technologies. The present TWM system in Sweden therefore leads to lower value generation from the used textiles. In order to extract more value however, there is a need to overcome the above-mentioned barriers, which logically demands a clear mapping and identification of the main actors and their interactions in this multi-actor system.

In this context "actors" refer to any institutional or individual entity that is engaged with the used textiles value chain, for generating tangible or intangible value for the TWM system. Even though the main actors of the Swedish TWM system have long been identified to be the charities, second-hand retailers, municipalities, fashion retailers, consumers, research institutes and academia, their role in the system concerns a wide range of activities including physical organization of reverse logistics of used textiles, new technology development, and knowledge creation and sharing. However, considering the highly fragmented nature of this system inherent due to confusion regarding ownership of waste or conflict of interest, the involved actors are in an urgent need for a collaborative dialogue (Watson et al. 2016; Palm 2011). To add, external factors such as lack of legislation and regulation further contribute towards this lack of collaboration in the system in undertaking extended responsibility for TWM (Pal 2016). Additionally, studies depict the absence of a well-functioning reverse logistics system for reuse and recycling of textiles in Sweden and consequently address the need for developing innovative structures and business models to enable higher value generation (Palm et al. 2014; Tojo et al. 2012). Clearer collaboration and joint responsibilities in such a well-functioning TWM system would provide incentives for operational improvements (collection, sorting and recycling) in the textile sector, while innovative business models would provide an opportunity for improving business triple bottom line, resourcefulness and lower environmental impact (Ellen MacArthur Foundation 2014; Palm et al. 2014). For instance, as highlighted by Bocken et al. (2013), elimination of issues related to ownership and financial risk taken through collaborative approaches among actors would lead to higher value generation and would provide opportunity for prolonging the active lifetime through reuse or recycling of used textiles (Tojo et al. 2012). Even though innovative business models such as leasing or renting offer an extended life-time for textiles and suggest collaboration among various partners, they hardly remove the challenges of a fragmented system as in essence they are single-actor business models rather than multi-actor (Palm et al. 2014).

In context to the above challenges inherent to the current fragmented TWM system in Sweden, a collaborative business model (CBM) would offer value and benefits to all the involved actors. The main goal of a CBM is to unlock barriers to sustainable innovation, for instance insufficient thinking out-of-the-box and unwillingness to work with external partners (Rohrbeck et al. 2009). The main concept behind CBM is to adapt the framework of business modelling as a tool for collaborative exploration of new markets and the planning of systemic innovations thus allowing the joint development of value generation system among the shareholders who can observe and reflect upon the problems of low value creation in a fragmented system with low levels of strategic collaboration (Ekström and Salomonson 2012; Rohrbeck et al. 2013). Thus the purpose of this paper is to understand the challenges to emerging opportunities of implementing a CBM by utilizing a value mapping approach, in context to Swedish TWM—an exemplar of a multi-actor ecosystem.

Given this purpose, our study clearly posits a novel perspective towards how to address toxic impacts of textile supply chains, thus directly contributing to the objective of the book. New TWM achieved through CBMs can lead towards higher degrees of reuse and waste prevention instead of textile waste incineration, thus directly inducing impacts in terms of climate change, photochemical ozone formation, eco-toxicity of freshwater, acidification, etc. than incineration and recycling (Schmidt et al. 2016; Zamani 2012). Reuse of clothes (even with a low substitution factor of 0.33), where substitution factor stands for how much is the purchase of new clothes being replaced by the sales of old clothes (Watson et al. 2014), can result in reducing climate change impact (in terms of person equivalents/ton) by almost five times compared to incineration (Schmidt et al. 2016), not to mention of the effects of substitution of progressively contaminating new production processes. Additionally, improving the value recovered from the reusable clothes is aimed at extending the life-cycle of existing clothes (addressing longevity), which has also proved to have positive environmental impact, e.g. 10% longer lifetime, i.e. ~ 3 months, results in saving carbon by 8%, water by 10%, and reducing waste by 9% (WRAP 2013). Hence, increased average lifetime of the used clothes is a decisive indicator of meeting the environmental objective as stated.

2 Business Model Literature Review

2.1 Business Model and the Value Concept

The notion of "business model" has increasingly emerged as one of the most used terms in management literature during the last two decades and was used commonly and constantly by business press and venture capitalists (Anderson 2016). Moreover, the importance of business models is undeniable as execution of a same idea or even technology will result in two different economic outcomes when it is taken to market through two different business models (Chesbrough 2010).

Even though the concept of business model is gaining traction in varied disciplines, it is still associated with being a vague concept which lacks a straightforward definition and compositional elements (Fielt 2014). Literature reveals varied definitions for business model presented by different scholars however it can be concluded that a business model is a representative of the value logic of an organization in terms of how it creates and captures value. Compositional elements such as customer, value proposition, organizational architecture (firm and network level) and economics dimensions, aid to specify the definition of business model (Fielt 2014). While academics and practitioners as mentioned above have not come to a conclusion on how to define business model, there are common grounds and shared views that they tend to commonly address for instance they all argue that a business model describes how business is carried out (Magretta 2002), and clarifies who the stakeholders and what their roles are besides, the value proposition for each of them is clarified (Timmers 1998).

A fundamental aspect of business models is how it reflects on value creation, exchange and capture logic from two perspectives, focal actors and business ecosystem (Chesbrough et al. 2006; Osterwalder and Pigneur 2002). Moreover, the viability of business model in some studies is considered to exist when all the stakeholders participating in that business model are able to capture sufficient value in a motivational manner (Chesbrough 2006). However, the notion of the term "value" within business model definition context is considered to be rather unclear.

Literature suggests that authors mostly have not been explicit and clear about the meaning of this term, meanwhile most definitions are limited in referring to customer value or value for the customer in other words (e.g. Afuah 2004; Dubosson-Torbay et al. 2002; Osterwalder and Pigneur 2010; Teece 2010). This lack of straightforward explanation for the term "value" has made it difficult to comprehend a definition of business model while a better understanding of the concept of value is not available (Anderson 2016). Despite the fact that a clear definition of "value" is unavailable, some authors such as Sweet (2001) identify four strategic value configuration logics which articulates a better understanding of this term. Value-adding, -extracting, -capturing and -creating are respectively the strategic configuration logics which add sustainable success to firms, and is achieved through the ability to manage these logics well rather than the ability to create new business model (Lund and Nielsen 2014).

In this context, some studies consider the value to be business models' conceptual focus, thus placing centrality to the interactions between the entity and other actors in the value domain (Weill and Vitale 2001). This proves that value definition in the business model context has to span over the boundary of a single ego-centric enterprise, and instead include a network of enterprises in the focus of the business model. The business model concept depicts a horizon wide enough to include a network of partners in which the success of the organization is closely tied to entity's relationship with other firms in the network (Lambert and Davidson 2013). Emerging new modes of business models like multi-sided platforms of value creation are deemed to be a threat to the traditional professions since the structure of organizing and value-realization has changed. Arguably, network organizations and social community based business models will probably require updated management system, therefore new conception of accountability, control and leadership and new sets of stakeholder tensions will consequently appear (Nielsen et al. 2014).

2.2 Network-Based Business Models and Value Network

The concepts of network and network analysis, which have received considerable attention in recent years, help to frame a good understanding of network-based business models. Despite significant attention given to the role of inter-organizational networks and their effect on success or failure of a relationship, limited attention has so far been given to the evolution of networks (Lund and Nielsen 2014). However, Batonda and Perry (2003) have introduced three schools for network evolution respectively: stage-theory, state-theory and joining theory. The stage-theory germinates into two main models, based on lifecycle and growth stages, and focus on gradual development of inter-firm networks through sequential stages over a period of time. State-theory is in the opposition to the sequential thoughts on which stage-theory is based, suggesting that actors in collaboration should move in a random manner from one state to another. Joining theory places its focus on the further development of the network (Thorelli 1986; Batonda and

Perry 2003). Companies new in the network settings most likely consider the collaborations as following a sequence of stages, while more established companies and also already network-based companies are likely to accept the state-theory approach (Batonda and Perry 2003).

Business networks of different kinds share various commonalities in terms of characterizations, such as being pictured as being formed by interdependent organizations which are in co-operation with each other and consisted of specific roles and value interactions (Heikkilä et al. 2014). They are oriented towards the achievement of a certain outcome (Allee 2008) to produce added value (Parolini 1999). A collaborative network is a network with joint processes in which partners share their information, resources and responsibilities to the plan, implement and evaluate activities towards the desirable common goal (Heikkilä et al. 2014). The overall aim of a collaborative network is mutual benefits for the involved stakeholders (Christopher et al. 2008). Such collaborations cannot be built based upon a contract, as trust is an essential requirement and enabler of co-creation. In case the trust does not exist, partners won't be willing to share their knowledge and ideas while this is a crucial aspect of business creation. Open communication and knowledge sharing are the ways to reach trust. Trust however, is not the only required element and enabler but also honesty, consistency and respect are needed as well (Larson and LaFasto 1989). By putting centrality to the value proposition in business networks, Heikkilä et al. (2014) has highlighted the triple role of a network's business modelling process, encompassing first, learning and knowledge sharing along with trust between parties, second, the presence of a formal coordination mechanism which shows the agreement over processes and rules, and third assessment of the risks, rewards and fairness of the deal.

As previously mentioned the concept of business model has the potential to span the boundary of a single enterprise to a network of enterprises (Zott et al. 2011), and it is through these potentials that a business model can enable a network of enterprises instead of a single enterprise to be the focus of a business model (Dahan et al. 2010). A network-based business model thus can be defined as the business model that at least two and often several stakeholders create with a joint value proposition depending on the key activities and resources of all the stakeholders (Christopher et al. 2008; Lund and Nielsen 2014). This clarifies that partners are not merely restricted to a traditional interaction along the value chain, but also can perform downstream customer activities and even core value proposition activities jointly. By doing so, Lund and Nielsen (2014) states that a network-based business model introduces competitive advantage by enabling companies to tap into and out of these networks, processes and inter-organizational relationships, and tap the ability to innovate across the network capabilities. However, network-based business models even though act as a hub for innovation which can accelerate the value creation and lead to development of global business models, there is a rather limited number of companies that practice and leverage innovative business models in networks (Lund and Nielsen 2014).

2.3 Collaborative Business Model (CBM) and Sustainable Innovation

The idea of CBM encourages the adaptation of a business model framework to be utilized as a tool to plan systemic innovation while having collaborative exploration of new markets. In addition, a CBM ought to encourage the development of joint value creation while exploring the benefits of mutual value capturing. In order to achieve this goal a CBM places it concentration on using methods which encourages and promotes creativity, decision making and planning—in addition to methods and processes that prepare a fertile ground for adapting the business model in collaborative manner (Rohrbeck et al. 2013). A CBM according to Rohrbeck et al. (2013) is a kind of business model that multiple organizations and actors work in an orchestrated style to create a value creation system or to jointly create the value capture system. These engaged actors can be diverse, for instance based upon their financial and legal status, i.e. for-profit vs. non-profit or public vs. private, based upon their position in the value chain (service or manufacturing), etc.

There are three main purposes for collaborative business modelling identified by Johnson and Suskewicz (2009); first is the need for the multiple actors to take actions in an orchestrated fashion so as to create complex systemic innovations which produce sufficient value to create a new market. Second, investigations show that there is a need for further coordination in cross-industrial areas as well as where the new technology is developing (Lei 2000). Lastly, to meet the requirement of developing new markets demands planning and decision-making, whereas there is still a high level of uncertainty in it (Ruff 2006). In such situations the process of decision-making can be facilitated by collaboration among different organizations in predicting the key developments (Rohrbeck et al. 2013).

In order to address the complex systemic nature of innovation—one of the key characteristics of sustainable innovation—multiple actors are required to work with each other in a coordinated manner. In this context, sustainable innovation could be defined as "innovation that improves sustainability performance", where such performance includes ecological, economic, and social criteria. Therefore, sustainable innovation will have different meanings and characteristics in different contexts and it goes beyond regular product and process innovation. In the case of sustainability challenges, the notion of innovation—in particular, sustainable innovation—is also connected to developing new business models (Porter and Kramer 2011). This demands new methods and approaches in order to jointly identify opportunities and plan sustainable innovations. Thus, it could be argued that unlocking challenges to (sustainable) innovation is the overall aim of a CBM (Rohrbeck et al. 2013). Examples of these challenges are insufficient thinking out-of-the-box as a result of being accustomed to daily routines, reluctance to change basic assumptions together with inadequate willingness to work with external partners and the lack of persistence in driving innovation (Rohrbeck et al. 2009).

2.4 Value Mapping Tool

In context to the wider notion of value proposition in CBMs across the network of stakeholders like suppliers, local communities as well as society and the environment, existing models such as the "business model canvas" (Osterwalder and Pigneur 2010) falls short due to their ego-centric perspectives on just capturing consumer value (Bocken et al. 2013). Therefore, to address this shortcoming network-centric tools can be used that take further values than only customers' into account. Tools such as "Value Network Analysis (VNA) (Allee 2011) or value framework introduced by Den Ouden (2012) even though offers a logical lens to understand the value in networks, sometimes these tools are too complicated and time-consuming, or are not specifically intended for business modelling, or could only focus on one dimension of sustainability for instance environment or financial value and thus do not manage to provide a holistic perspective (Bocken et al. 2013). Additionally, another issue associated with the current tools is that they are either only conceptual or not used in industries widely and rely on an external well-trained facilitator. Hence, a need for a tool which assists the firms to develop a better understanding of the sustainable value creation within the context of their business activities, plus, assist them in developing new business models which have the sustainability as their core value, is the need of the hour.

"Value Mapping Tool" is a tool that tends to go beyond "egocentrism" and economic value, and provides a method to develop shared value propositions for stakeholders, as diverse as customers, suppliers and also governments. Such a network-centric rather than firm-centric perspective of this tool leads to the optimisation of value within a network. Through differentiated mapping of the value captured, destroyed, missed and new value opportunities (Bocken et al. 2013), this tool is aimed at assisting firms to create value propositions that are better compatible with sustainability. Four major stakeholder groups which are environment, society, customer and network of actors, are being assessed through this tool for better understanding of their value proposition, considering both positive and negative aspects for the participating stakeholders in the value network. Thus it offers a unique and novel way of conceptualizing value by adapting multiple stakeholders' idea of value and a network perspective as opposed to single firm-centric perspective.

In Fig. 1, the conceptual portfolio of value innovation opportunities for a firm as well as its involved stakeholders are illustrated. The value proposition of the network is placed in the core of this portfolio through which the benefits which are delivered to stakeholders are represented. Stakeholders achieve these values through payment or another value exchange. It is to say that, in the process of delivering the value proposition, it is possible that the individual stakeholders and networks destroy value. This can occur in various forms however it is worth mentioning that destruction of value in the context of sustainability is mainly concerned with environmental damages and social impacts of the business activities. Bocken et al. (2013) refer to value destroyed, as "negative externalities" however it is argued that

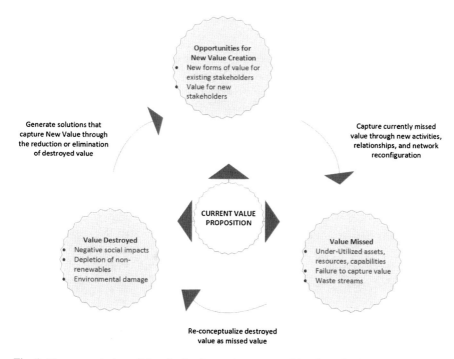

Fig. 1 The conceptual portfolio of value innovation opportunities (from Bocken et al. 2013)

this might not be the most correct terminology as it seems to be distanced from the firm. Missed value opportunities are referred to the situations in which stakeholders do not succeed in taking enough advantage of the existing assets, resources and also capabilities operating and working under the industry's best practice. Similarly, missed value is referred to the time when a firm does not receive the benefits which it seeks to obtain from the network. The participating reasons for this can be a weakly designed value creation or capture system, failing to persuade others to pay for the benefits, or failing to acknowledge a value. Expansion of the business to a new market and introduction of new products and services that offer benefits to the stakeholders are known as the "New value opportunities". This can go beyond the customers and address values such as employee well-being or positive contributions of the environment. A comprehensive and thorough exploration of a business model and also identification of areas in need of change and improvement can be resulted through the identification of these values.

Moreover, value mapping tool adopts a qualitative approach to value analysis. In the area of sustainability utilizing quantitative analytical tools such as "life cycle assessment" are common, despite that, the proposed use of the value mapping tool at the business model level is not concerned with the quantitative details as it is related to qualitative details such as stimulation and generation of idea and discussion (Allwood et al. 2008).

3   Methodology

This study is conducted through a series of semi-structure interviews and a selective approach to representative organizations. This is accompanied with inductive reasoning to investigate the relationship between theory and research. Qualitative data in this manner is produced in the form of detailed answers that the interviewees provided to the open-ended questions. The approach for this qualitative study is most suitable and is classified as exploratory since it attempts to investigate a fairly unknown area (Wiid and Diggines 2009). Recognition of the seven major actors, viz. charities, fashion retailers, municipalities, research institutes, academia, private collectors and consumers, occurred first during a pre-study. In the pre-study, two mappings of: "main actors" and "actors' activities" were made in Swedish TWM system. These mappings later helped the interviewees for better visualization of the TWM actors and activities. The "main actor mapping" aided the interviewees to recognize who the actors were organized in the system, which led to identification of possible missing actors. This was followed by activity mapping of the seven main actors based upon the EU waste hierarchy (see Palm 2011), to explore the possible interactions among the actors.

The nature of this study which was mainly focused on mapping the actors, their activities and interactions required the authors to conduct interviews with selective representatives from organizations in Swedish TWM system who already have the most engagement or have plans for extending such engagements in future. The group of "consumers" however was an exception since despite their important role in the TWM system they do not form an organization which acts according to a specific business model, but rather they are individuals. Since the main focus of the study was on the business models, main goals and values of the organizations, the consumers were exempted from interview however their role and impact on the system has been mentioned in the study.

Data collection, as mentioned above, was through semi-structured interviews (Yin 2014). The authors framed open-ended questions in a manner that enabled the respondents to reflect on the identified activities as well as their existing or future collaborations with other actors. Seven interviews were conducted in total, out of which six were in person and at the respondent's facilities, while the remaining one was over phone due to time limitation on the respondent's side. This resulted in a total of 6 h and 30 min of recorded interviews. Subsequently, the data was analysed by the aid of the value mapping tool (developed in Bocken et al. 2013), where the core values (social, environmental and economic values) were mapped based on actors' activities and interactions. At the same time, the areas where actors had common or conflicting values were revealed. Furthermore, authors applied coding scheme based on the value innovation portfolio (shared, missed, new opportunity, and destroyed values) to analyse the interviews transcripts.

4 Findings

In order to conduct actor and activity mapping we have followed the EU waste hierarchy (Palm 2011), i.e. prevention, reuse, materials recovery, energy recovery and landfill. Certain adaptation has been made during the mapping to suit the waste hierarchy to the existing Swedish TWM system. For instance, due to the lack of large-scale recycling in the Sweden, the key actors considered are academia and research institutes for their involvement in developing knowledge and practices related to textile recycling and activities such as generating new technology.

4.1 Mapping of Actors in Swedish TWM System

In this section, first a brief introduction of the actors and their organizational activities is presented in the process of actor mapping. This is followed by presenting the outcomes of the activity mapping analysed in line with the value mapping tool. The section concludes by presenting what the respondents perceived to be the key enablers for collaborations and networking for CBMs in the Swedish TWM system. Seven main actor types were recognized to be operating in the Swedish TWM system; these are charities, fashion retailers, municipalities, research institutes, academia, private collectors and consumers. In addition, government institutions enforcing environmental regulations, such as the Swedish Environmental Protection Agency (Naturvårdsverket), was also considered important for the system for setting goals, strategies and rules that can affect other actors' activities and strategies.

4.1.1 Charities

The charities are traditionally the largest collectors of used textiles in Sweden. Their collection takes place in different manners; as mentioned previously they are allowed to place their collection boxes in the municipality areas and recycling centres, while collection also takes place through bins in their own second hand shops. Some charities also offer services such as "call & pick up" for donors who have difficulties in bringing their donations to them. Sometimes fashion retailers also provide the charity organizations with their "waste" such as defective pieces or leftovers. The used textiles collected by the charities are destined to second-hand stores, while some are being reworked in own remake and redesign studios. Revenue is generated by the charities by selling through the second-hand shops or by exporting them. These activities are supposed to add both social and environmental values. At times when the collected textiles are not in a condition to being either sold in second-hand shops or being exported to other countries, the charities send them for incineration. These are typically considered industrial waste and thus required to be paid for incineration. However, incineration plants often exempt the charities of such charges due to their format as social enterprise and also good will.

4.1.2 Fashion Retailers

Fashion retailers (including second-hand retailers) are a group of actors who increasingly influence the flow of used textiles. They collect textile waste either through their in-store collection boxes or with the aid of private collectors for varied purposes, such as second-hand sales, exports, redesign, or incineration. Additionally, some fashion retailers (and second-hand retailers) store their collected used textiles in own stock to have full control over their value chain also due to limited volume and lack of recycling technology. Sometimes the fashion retailers also donate leftovers and dead stocks to charities as humanitarian operations. Recently the retailers have also started to collaborate with commercial sorting firms who then redistributes or exports them to developing nations. Finally, when incinerated, the fashion retailers have to bear a fee since their textile waste is classified to be industrial.

4.1.3 Municipality

The municipalities are the main actors responsible for environmental waste treatment. When it comes to TWM in Sweden, there is no separate collection system for textile wastes and thus used textiles are a part of either household waste collection or industrial waste collection process in most of the municipality recycling centres. Some municipalities give the permission to charities holding the 90account certification (offered by Swedish Fundraising Control) to put their textile containers in municipality areas. The municipalities are also in charge of sending some portion of the collected waste to the incineration centres for which they have to pay a fee.

4.1.4 Research Institutes

Research institutes play a significant role in the development of new technology as well as spreading scientific awareness to the public about TWM systems in Sweden. They play their part by educating the public through publishing the result of their investigations. In some cases research institutes and academia work together in finding a way towards developing new technologies. Both the actors, research institutes and academia, conduct their research with the aid of large or small organizations and fashion retailers who are willing to invest heavily in developing new recycling technologies.

4.1.5 Academia

The role of academia is to, together with their stakeholders, identify the "needs" in the textile industry and educate students and future researchers to address them. Academic activities are not a business proposition, but rather a task proposition. However, their task does not include taking actual actions, but is focused on providing inspiration, information and facilitating other actors. This way academia builds empirical material to be utilized further by their staffs and researchers. Such initiatives are co-funded by other actors, for instance, the region and organizations.

4.1.6 Private Collectors

Municipalities are capable of collecting only around 30% of the waste in Sweden and for the rest they hire sub-contractors and private collectors to take care. Big private collectors are often the ones in collaboration with municipalities however they do not have any specific fraction for the textile waste, thus their collected textile material might be mixed with other kind of wastes. As the private collectors of used textiles cannot have the 90account certification and thus be allowed to place their collection bins in the municipality areas, the often team up with the charities to gain access to wider collecting points.

4.1.7 Consumers

The consumers are the main decision makers regarding the destination of the textile waste. Consumes are responsible for doing the initial sorting of their textiles in wardrobe, to decide what to do with their household used textiles. and are connected to other actors such as charities through their donations of used textiles. Their role is not only limited to disposal of the used textiles, but they actively influence the actions of other actors as well, for instance consumer attitude and awareness can affect the fashion retailers' long term strategies to use recycled material in collection or utilization of new technology or their engagement with TWM system to take extended responsibility.

4.2 Activity Mapping in Relation to Actors' Interactions

This section highlights the mapped activities and collaborations of the engaged actors. The value mapping tool is applied to reveal and analyse the areas in which the actors have shared or missed/destroyed values and also for identifying new opportunities for value creation. Utilizing the tool, actors' activities and collaborations were fist assessed which resulted in better understanding of actors' core values. Further it introduces the current value proposition of the actors, value destroyed and new opportunities for value creation. This eventually articulates to a better realization of the potentials for their collaboration in a form of CBM for sustainable innovation in TWM and its possible impact on Swedish TWM system.

4.2.1 Collection

Charities, municipalities, fashion retailers and private collectors are the actors who play an integral role in collection of textile waste in Sweden. Their collaborations along with the shared and/or destroyed/missed values are narrated below, which lays the foundation for investigating the new opportunities for future collaborations.

Through charities

Charities being the largest collectors of textile waste in Sweden, in some cases, collect used textiles through collaboration with private collectors, even though these private collectors rarely collaborate with the municipalities for collection of textiles. Large private collectors of solid waste Sweden, such as SITA and Rangsells hardly find the motivation to collaborate with municipalities in collecting textiles. This is because collection boxes contain only small portion of textiles material in comparison with other kinds of solid waste. In addition, there is no significant financial incentive for the private collectors to invest in collection of textile waste resulting in "value missed/destroyed". However, as stated:

> incentive for big private collectors will appear in case municipalities due to procurements related to textiles decide on hiring the private collectors for this task. (Municipality)

This created an opportunity for generating shared values. The municipality representative further explained that the main difference for initiating collaboration with charities and private collectors lies in the motivational drive; while the former is by its social values the latter is by economic values. In other words, private collectors tend to run a business which is environmentally beneficial. Private collectors on the other hand highlighted the benefits brought through collaboration with the charities, by gaining access to exporting channels. In addition, this also results in more reliability in the eyes of the public and assures the private collectors corporate social responsibility (CSR) promises. In such collaborations, charities benefit economically as it is cheaper to operate through large private collectors for organizing collection services instead of operating alone in installing and maintaining collection boxes on municipality grounds (for which they have to pay a certain fee). In addition, collaboration helps the charities in saving reverse logistics costs—an answer to the dilemma posed by on one hand demands for logistic cost reduction and on the other increased logistics due to upscaling requirements. Thus the private collectors by providing the collection and reverse logistics services can aid the charities to move towards their social goals. Moreover, the collaboration also aids the charities to have an access to different markets and collection points via the private collectors, thus creating new opportunities for generating shared values between the charities and private collectors.

Through municipalities

As of date, municipalities mainly operate their used textile collection via charities, as the charities places their collection bins at the municipalities' recycling centres. Direct collaborations with collectors and sorters of used textiles either inside or outside Sweden might be their new approach in the upcoming future thus resulting in new opportunities. The current system however incorporates certain challenges and shortcomings resulting in value missed/destroyed. Lack of a unified collection system for textile waste in Sweden due to differing collection schemes in different cities have posed a major challenge to the municipalities. This difference in collection schemes and systems might be due to different waste management companies that municipalities are in collaboration with. These waste management companies typically operate according to different legislations, however due to material-wise difference in the legislation has posed a major problem in organizing these service, thus resulting in value missed/destroyed. For example, material such as glass has a definitive legislation which should be followed unlike textiles which lacks one.

New streams of collection owned by municipalities are also being progressively developed. For instance, in cities such as Stockholm and Malmö, for collection of textile products like carpet, sheets and households a new category called "other textiles" has been introduced. This new category may result in less incineration of textiles and is a step towards more environmentally friendly treatments, thus opening up new opportunities. According to a test conducted by the municipality in one of their most visited recycling centres, up to 4% of the mixed waste collected was textile waste (a relatively high percentage) which currently has no other option left other than incineration. As municipalities are required to pay for the incineration of the collected waste, separation of textile material from other kind of waste will result in large saving on incineration costs and provide opportunity for greener treatment by reducing the carbon footprint of their actions. Moreover, identifying the right collection bin in the municipality-led recycling centres can be sometimes confusing for the consumers which can eventually lead to missing value potential. One of the charities stated that the municipalities can address this issue by considering a certification or a label to the charities for their sustainable and environmentally friendly actions and are working under their standards and principles. Municipalities for instance, have already introduced such schemes, such as the Golden Certificate (in Swedish Guldmärket) which is a label given to those organizations who are involved in reuse and reduction of waste, thus offering new value creation opportunity. In extension, 90account is another example of a certification scheme for social enterprises, aimed at ensuring credibility of the collector in the eyes of the consumer.

Furthermore, flea-market is a new phenomenon which shows a different kind of collaboration between municipalities and charities. High amount of textiles are being exchanged among people in these markets and at the end a massive portion is left in the area. Municipalities in such cases tend to call charities to collect the leftovers. However, due to lack of sound reverse logistics and storage the charities often face the challenge of handling such large volume of textiles in one single day, resulting in missing economic value creation potential. In this context, the respondent from the charities reflected upon development of tighter collaboration among the charities to support more efficient collection, potentially generating higher shared value.

Through fashion retailers

More recently, collaboration between private collectors and fashion retailers have emerged in the TWM arena. Private collectors arrange a way to send the fashion retailers' collected textile waste to the charities rather than for incineration. Considering the relatively good quality of these textile materials, they can be of good use in the second-hand market for the charities. However, the challenge is the locational decision related to the second-hand market where these used textiles are sold, as the fashion retailers are sensitive about the market that their products end up in. This addresses a "value destroyed/missed" among charities and fashion retailers.

Additionally, another sensitive aspect in such relationships between the retailers and the charities is the pricing of the defective or dead stocks. As the price of these items is relatively low in the second-hand market, this concerns the fashion brands of relatively higher prices.

> [...] consumers don't really care if it's the old collection, they look at the brand. So, they would see a very high-end brand in our shops and they see the clothes are new, then perhaps they prefer to buy it at our shops because it would be less expensive so that would make for them a big competition [...]. (Charity)

However, as added by the charities new legislation regarding producer responsibility can smoothen the collaboration between the two actors. This suggests a possibility to transform the "value destroyed/missed" to "new opportunities". On the other hand such initiatives can provide immense help to the fashion retailers to better respond to their CSR promises and show extended environmental responsibility and corporate citizenship. In addition, this leads to saving costs associated with logistics and incineration fee. This suggests both "new opportunity" for "shared values" among charities, private collectors and fashion retailers.

Through emerging collection platforms

New emerging platforms such as mobile applications (e.g. Cirqle app) can introduce a new channel, and consequently a new market for charities and fashion retailers for the collection of textiles. Collaboration with these emerging platforms can also provide good reverse logistics support that charities often have to wrestle for. At the same time, it will encourage further participation of the donors, thus providing new opportunities by overcoming the value missed due to lack of sufficient reverse logistics facilities. As explained:

> [...] so each time someone decides to donate clothes, they first sort all their clothes in their wardrobe, and then they decide what to give away, next they can use the app in which there is a map that shows nearest donation points where they can leave their donations, it's only charities [...] for example, they would go to Stockholm city mission, they would leave their

bag there and after that they would go to the cash register to show a code, then they will get 20% of reduction for example on some new products like happy socks […]. (Charity)

Such innovative collaboration schemes are foreseen to accelerate their shared values towards more efficient and environmentally friendly collection system.

4.2.2 Reuse

The term "reuse" in this study refers to activities such as resell in second-hand stores, garment redesign, donation, and exports. Export is included in this classification as the purpose behind it is to further sort the used textiles for sales in mostly developing country markets "as they are" or sometimes are converted into mechanically recycled products. In other words, these activities expand the reuse of textiles in different forms. Textiles with higher quality, known as *prime*, are often kept in the country of collection (here, Sweden) to be either sold in domestic second-hand stores or are redesigned. Out of the remaining, approximately 10% of the used textiles (possessing nearly 50% of the sales value) are being reused mostly in Eastern Europe—often in the same countries where they are sorted. The next quality grades A and B and tropical quality tend to be exported to Africa, the Middle East and Latin America. Industrial wipes and the lowest grades are exported to Asia for mechanical recycling purposes. India and Thailand have significantly large industries for shredding and unravelling these textiles for use in production of low grade textile products, such as rugs.

As second-hand and redesign

The charities are widely involved in reuse activities, from second-hand sales to remake and redesign and in some cases direct donation for humanitarian reliefs. These activities are organized under a social enterprise with social purposes. Even though remake studios and second-hand stores both serve the same purpose for the charities, sometimes these channels tackle competitive clashes. For instance, fashion retailers of high-end brands mostly prefer to donate the clothes to the remake studios as they would prefer to have their products being sold under a different brand and with slightly different design rather than at a lower price in a second-hand store. While, on the other hand charities are constantly looking for products of higher quality to be sold in their second-hand stores as their customers look for affordable good quality clothes. This exemplifies a case when value is destroyed due to the inherent conflict.

Also fashion retailers are active in the reuse area through their own second-hand stores and remake studios, and as stated by one of the respondents from the research institute can be foreseen to be growing if a mandatory producer responsibility is put into place. As fashion retailers run their own second-hand stores the actively reflects their social engagement and responsibilities. However, implementation of these activities is to some extent dependent on fashion retailers' size and strategies. Presently such activities are on a small scale, and most of the fashion retailers only manage through collaboration with other actors in the TWM system, such as charities and academics.

As exports

Charities are actively involved in the export of used textiles, sometimes with the collaboration with other actors like municipalities or private collectors. This collaboration benefits the private collectors by providing access to the exporting channels of charities, which suggests a "shared value" between actors. Municipalities do not export used textiles on their own, however they recently have had a pilot collaboration test with charities to do so. This collaboration suggests a "shared value" that enables the "new opportunities" for new activities. In addition, there are examples of their collaboration with partner companies who are involved in exporting textile waste. For instance, municipalities' contract partners send textile waste to global sorters like KICI or SOEX in Germany and Holland.

Fashion retailers' also engage indirectly in exporting their collected used textiles, as they collaborate with private collectors in exporting to other countries, which is either resold as second-hand items or are recycled. The collaboration of I:CO with fashion retailers can be mentioned as an example of this kind of collaboration. Fashion retailers however do not export through charities.

However, several factors such as decrease in global fibre consumption in the near future and the development of the traditional export markets, like Africa due to growth in national production, forecasts a decline in the demand for used textiles. Consequently, this may result in a decrease of used textile exports, thus suggesting a "value missed". Therefore it could be argued that the necessity of switching to other options is undeniable which offers "new opportunities" to address "shared value" between actors who are involved with used textiles exports.

4.2.3   Energy Recovery (Incineration)

In Sweden the incineration plants are owned by municipalities who receive a fee as an exchange of the service they provide for the actors. Charities, fashion retailers and also private collectors who are in one way or another involved in collecting used textiles sometimes send some portion of their collected textile waste to the municipality-owned incineration centres.

In general, incineration due to the high negative impacts on the environment is considered to be a controversial waste treatment method, which is worsened when it comes to textiles. This is because textiles are not desirable fuel for incineration centres as they do not burn properly and it is difficult to have control on its process. Additionally it brings high environmental costs although it comes in small portions therefore it is not an attractive fuel for incineration centres. This point out to the potential of generating a "new opportunity" for the involved actors towards collaborating for a better treatment of textile waste, thus generate higher "shared value".

4.2.4   Recycling

Textile recycling in Sweden is limited to small testing scales therefore this section is limited to describing activities related to textile recycling technology development, as undertaken by the engaged actors.

Generally the research institutes collaborate with all other actors in an attempt to develop new technology, thus highlighting their aim towards "shared value" creation through such collaboration. As stated:

> […] with the fashion retailers the relationship is that we develop new technology for them, and our relation with municipalities is that the new technology does not interfere with any regulation, and the connection that we have with charities is to try to see where their feed stock would fit to the industry. And with academia we collaborate as they are more into basic research compared to us […]. (Research Institute)

Several dyadic collaborations can be pointed out in this context. As mentioned above, municipalities collaborate with the research institutes. Although such collaboration in TWM system is rather new and limited, it is an attractive area for the municipalities to increase the textile recycling potentials. At the moment their collaboration is mainly related to joint projects with the aim of knowledge sharing about the technology developments and gaining a better understanding of the textile flow system.

Charities too engage in collaboration with academia and research institutes. This collaboration can be in form of participating in research projects by providing used textile samples for conducting experiments. Further information sharing (e.g. statistics on volumes and fractions) is also a key aspect of such collaboration as the charities depend on this form of information to further share that with the end consumers. Charities on the other hand can act as a bridge between customers and research to pass important information from the customers, such as consumer acceptance level, etc., and vice versa, thus reflecting the potential to generate "shared values" among the actors.

Private collectors also show interest for being in collaboration with research institutes, for instance participating in pilot projects. This is crucial for private collectors to learn about the new demand of the market, latest technologies, etc. in order to have a better approach to their customers. This generates "new opportunities" for "shared value" creation among the actors to reduce the risks associated with the market. Even though collaborating for related technological advancements can be perceived as a potential to change the landscape of TWM in Sweden, there are several challenges to it. For instance, new technology development is a time consuming process to make it market ready commercially; also the scale of the current testbeds are relatively small. Private collector interviewee highlighted the important role of science parks and incubators in development of new technologies in Sweden as they have the capability to realize the demand and make the new technology to a realistic tool for commercial use for different actors in the system. This may provide a "new opportunity" for facilitating development of new technology.

4.3 CBM for Sustainable Innovation in TWM

4.3.1 Actors' Shared Values

This section highlights the main motivations for the actors to participate in a CBM for sustainable innovation.

The fashion retailer highlighted that one of the main drivers and motivations for their participation in developing CBMs in the textile waste management system is related to corporate social responsibility (CSR). On the other hand, for the research institute and academia collaboration with other actors pave their path to have a better introduction of the newly developed technology and innovations. This further enables them to expand their studies by building upon a strong empirical knowledge base. Further this also stimulates natural research; for instance in collaboration between academia and municipalities the situation naturally becomes ready to take real actions along with investigations. This is beneficial to municipalities as they are eager on collaborating by making long lasting investments that can eventually benefit the citizens.

Similarly, as stated by one of the charities, they also welcome innovative alternatives which enables them to spend their budget more effectively, thus collaboration can help them to overcome the challenges related to time and costs. For instance, a more efficient sorting technology in this manner makes a considerable change in their working process. Such technology would similarly aid private collectors as well. Such collaboration is of equal importance for the municipalities, and as stated:

> [...] receive varied types of solid waste which should to be treated efficiently. Textiles material is only small portion of the waste we receive; therefore in order to organize an efficient recycling for this kind of material there is a need to work in a network with other actors who are involved in TWM scenario. (Municipality)

4.3.2 Actors' Missed/Destroyed Values Due to Conflicts

As previously mentioned, one of the main challenges to collaborate is the different goals and values of the actors, which hindered shared values identification. To highlight it was said:

> [...] this business model could hardly involve everyone business models. Their entire goals could be matched but it is possible for just some of them in this collaborative business model because of some conflicting values [...]. (Private collector)

Research institute interviewee stated that there are examples of collaboration among the actors at the moment which is mainly in forms of knowledge sharing however, collaboration in a form of a business can reduce the tendency of the actors to participate and merely those actors who can seek financial benefits would participate. Municipality's representative similarly argued that financial aspects of collaboration tends to complicate it, particularly when the result of negotiations for the possible collaborations is close to implementation as this is the time that new challenges pop up.

Conservatism and normativity are the two main challenges that hinder the actors to collaborate. Actors need to be open to develop new values and not to be myopic and short-sighted in their works.

> […] it's not a local problem, it's a global problem. The main value has been very much concentrated on financial value. So, if you want to embrace this new world you have to have some other values […]. (Academia)

In order to have a successful collaboration it is crucial for the actors to have a comprehensive understanding of the nature and essence of each other's business values. Lack of this understanding can raise challenges in actors' collaboration. As an example, one of the respondents from the charity mentioned about the possible challenges of collaboration between charities and research institutes arising due to different business values despite their common interest in innovation.

> […] the research institute are interested in innovation and we are too … if they would collaborate with charities I'm not sure they would understand our situation because working out here with people is so different than their work… so, it would be really great if we could understand each other in some ways, they could come over here and see how it is here and.. but it takes time and investment for us also as charities […]. (Charity)

4.3.3 Actors' Opportunities for New Value Creation

Keeping in mind the notion of CBM as a solution for TWM scenario, identifying opportunities for new value creation is essential to turn the missed/destroyed values due to potential conflicts into shared values. Most of the interviewees were in agreement to the fact that the actors need to collaborate in order to operate an efficient TWM system, as none of them can single-handedly manage it. It was evident that the lack of resources and capabilities of one actor can be complemented by another actor, to generate new values collaboratively.

> […] fashion retailers can't do it alone as they don't have expertise but they have money, we can't do it as we don't have money and the volume, municipalities cannot do it themselves as they need fashion retailers and organizations like us so the fact that it has to be collaborative is definitely right and the answer is absolutely yes […]. (Research Institute)

Additionally, all the actors confirmed unanimously that there is a lack of proper communication among them in terms of TWM. Hence it would be valuable, as suggested by the Research institute representative, to initiate a platform for the actors to communicate and come together.

Nevertheless, the issue which needs further investigation when actors' collaboration is supposed to be in the form of a business mode is their core values. As was highlighted by one of the interviewees, the core values of each actor need to be boiled down in order to reach a shared value among them. It was argued that it is important how different actors are set up in a CBM, however may be not all of the actors can be fit in one single business model, but there could be an umbrella business model as a platform to feed and serve other business models and result in their development. Reflecting along this thought:

> [...] this is collaboration but each one is individual... and you collaborate on the things which are not negatively affecting your business. A collaborative idea is an advantage for the business and it provides means for better pricing, better fractions for recycling, which cannot be generated by each one [...]. (Academia)

4.3.4 Key Enablers of Collaboration and Networking Among Actors

The role of legislation, such as producer responsibility, was highlighted as one of the main drivers to make all the actors be on-board to take the financial and organizational risks of collaboration. This would help them to serve their extended responsibility in a better manner by using each other's competence and capabilities. As it was highlighted:

> [...] to get actors in different fields to take little bit of risk may not happen until coming of new legislation. That could be the top point which is coming soon and that is why they are all recently so active and willing to take the risk, otherwise they will have a lot of waste material which cost a lot to send to incineration [...]. (Research Institute)

A proper legislation was perceived to be leading to quicker progress in TWM system, and possibly overcoming its challenges however, there is a requirement for hand-in-hand development of new textile recycling technologies since a new legislation will require the actors to handle their textile waste effectively and this will only happen when the required technology is developed.

In addition, actors unanimously identified the need to define a clear structure of CBM to be one of the main enablers. A CBM has to be market-oriented and purposeful in way that market defines the required collaboration and the actors who should be collaborating with each other. Consequently, the collaborative landscape which is being proposed should avoid having a fixed structure, and instead should be flexible to help flourish new CBMs within itself. This will provide the actors an interactive platform in which they can identify collaboration areas that they missed or have been unexplored so far. Therefore, the drivers of a CBM are dynamic and are subjected to undergo constant change all the time. As was highlighted:

> [...] the form of collaboration among them will change in a different period of time as something that we develop now which is high tech now in 30 years that will be commodities (old) and then people will try to develop something else [...]. (Research Institute)

In this context having smoother dialogue and building of higher trust among the actors were highlighted to be essential. Setting up a neutral platform in which the actors can freely discuss their core values and concerns was highlighted and in such case many actors identified the role of academia to play such a neutral role among all the actors, thus avoiding any conflict of interest.

5 Discussion

5.1 Value Overlaps in Ego-centric Business Models in Multi-actor System

The Swedish TWM system is a multi-actor system, as it is regulated by varied organizations, companies and channels. Each actor has a certain business model with set of competences through which its approach to and its functions in the TWM scenario is defined. Underlying each business model is a different set of values and focus, rationalized by actors' core activities and responsibilities. For instance, the municipalities, with the aid of Swedish waste management companies, are responsible to provide "service" to the society by providing environmental waste treatment for the household waste. This service-based task makes the municipalities' business model to be distinguished from other actors' as each organization depending on the essence and nature of their business develop their business model (Boulton et al. 1997), to combine their assets in order to create value.

Thus each actor's activity is based on an ego-centric business model, and it is evident that most of the actors prefer organizing their operations for TWM in their own way. Even though in such ego-centric business approach the actors concentrate on their individual goals and values, often these values overlap among the actors as was seen in case of TWM system, thus addressing the potential of individual business model concepts to span over the boundary of a single ego-centric enterprise and instead include a network of enterprises into focus (Lambert and Davidson 2013).

5.2 Collaborative Activities in Multi-actor Ecosystem

The multi-actor TWM system in Sweden also addresses the need for collaboration among the actors to complement each other's missed/destroyed values. Such a system of interdependent activities transcends the focal firm and spans its boundaries, thus enabling the firm in concert with its partners, to create value and also to appropriate a share of that value (Zott and Amit 2010), as was evident clearly at many instances in the Swedish TWM system. For instance, the charities and retailers utilized the logistics facilities of the private collectors, while the collectors in order to export their collected textiles utilized the charities' exporting channels. Heikkilä et al. (2014) states that such a collaborative network organizes joint processes in which the partners share their information, resources and responsibilities to plan, implement and evaluate activities towards a desirable common goal. The overall aim of such collaborative network is to achieve mutual benefits for the involved stakeholders (Christopher et al. 2008).

5.3   Challenges to Collaborative Activities in the Current System

Despite the need of the actors to work as a collaborative network certain obstacles are identified. For instance, lack of legislation and as a result confusion regarding the responsibility of the actors regarding ownership of the textile waste (Naturvårdsverket 2013; Palm 2011), which has further resulted in a fragmented system where the actors increasingly shift their responsibility to each other. This has resulted in lower transparency in the system towards defining roles and extended responsibility of the actors (Pal 2016), compared to other sectors like plastic and glass where the legislation is strongly in place.

Another key challenge is the existence of a varying working system for TWM companies as a result of lack of legislation. This increases the inherent confusion in the actors about their role and responsibility in a network thus motivating more ego-centric mind set hence business model. This further leads to a value loss as a large amount of textile items are still sent for incineration, which otherwise could have been subjected to higher value-adding recovery options.

Lack of a neutral platform for accommodating a purposeful dialogue among the actors is also a key challenge, which results in missing the opportunity to interact and freely discuss and expand over each actor's core values and concerns. Through our study, it could be highlighted that such a neutral communication platform organized by academic institutions as actors is favoured by the other involved actors. Altogether these have resulted in substantial loss of value among the actors, which can be addressed in a collaborative network. For instance, financial incentives for private collectors can be a good motivator to invest in collection of textile waste however this is presently quite small, but can be promoted by municipalities as they can seek the services of private collectors for collection of textiles. In yet another case, collaboration between the charities and municipalities for the placement of charity-owned collection boxes in municipalities' grounds can helps both the actors in taking a joint effort and moreover reduce fragmented collection. Such collaboration is also evident when the municipalities seek the help from charities in collecting leftovers from flea-markets.

5.4   From Collaborative Activities to CBM for Sustainable Innovation

Exploring new potentials for innovation and cross-companies collaboration on the basis of shared goals and normative values can be achieved by developing a better understanding of systemic innovation. Such systemic innovation can take place either in a form of a business model or in terms of sustainable innovative technology. Both can motivate the actors to shift from being single-actor with ego-centric view to engage in multi-actor collaboration, thus resulting in "shared-value" creation (Porter and Kramer 2011). For instance, in the TWM scenario, new technology in either sorting or recycling demonstrates to a certain extent how actors can collaborate with each other by seeking for a new technology and innovation which meets their core values and can lead them forward to a sustainable innovative business solution. More innovative collection systems developed collaboratively by different actors in sorting, reuse and recycling of textile waste can inspire further collaboration as well.

While the new technology and innovation can affect the components of an existing business model and how they interact, it can also create completely new business opportunities in uncontested market space (Schneider and Spieth 2013; Breuer and Lüdeke-Freund 2014). Such a role of CBMs in creating systemic innovation by addressing new markets is suggested in Johnson and Suskewicz (2009). In context to the Swedish TWM system, such innovative CBMs is set to extend the lifetime of the textile products through reuse or recycling and has the potential to address the existing or emerging issues in the market. In other words, the CBM can be deemed as a tool to plan the sustainable innovation, as it helps in developing a joint value creation system and at the same time explores mutual benefits of value capture system (Rohrbeck et al. 2013).

Sustainable innovation however is not always easy to promote and it has its own challenges as well, for instance when actors are used to their daily routines and are unwilling to change their basic assumption. Similarly, their unwillingness to work with external partners or when they are inconsistent in driving innovation, moving towards sustainable innovation can be difficult. This can be seen in the TWM scenario when the actors show unwillingness to invest in the development of new technologies due to their conservatism or due to low interest in changing their daily routines and go beyond the old system, as was highlighted in the study by one of the interviewees. Shifting from the present settings for waste treatment to a new method was reflected upon as time consuming. The issue of ownership of the new settings was also mentioned as a key challenge to such transformation. Further examples of actors' unwillingness to change can be seen in the charities and their concerns related to development of new technology. Despite their positive view to development of new technology, they have the concern that innovative new technology in areas such as sorting and recycling might also lead to fewer job opportunities in their social enterprises. However, research institutes as developers of new technological solutions proposed a different angle to look at its outcome, and mentioned that new technology would instead result in job creation, and that other actors concern on job loss due to technological advancements can be due to their organizational inertia and trust in the old system. In this context, in line with Breuer and Lüdeke-Freund (2014) we see a conflict in normative orientations of the engaged actors due to varying corporate visions and missions, and core business value, which poses a wicked challenge to achieve sustainable innovation. As mentioned by one of the interviewees, to bring in sustainable innovation it is necessary to go beyond myopic visions and instead be eager at developing new cross-industry values. Such adaptation is necessary for the consumers as well, as this means a change of what they are used to.

Considering the barriers of CBM for sustainable innovation, and the changing dynamic environment that a CBM is functioning in, it is crucial that CBMs should be customized and adapted to the environment according to the needs and requirement. Involved actor's critiques and analysis of a CBM could help to formulate and design a business model which can better respond and meet their goal and purposes.

6 Towards a Collaborative Business Model Solution

6.1 Conclusion

Through our study, three key drivers of CBMs emerged in context to the TWM system which can confront the challenges arising due to organizational unwillingness to change, financial risks related to new advancements in this system, and lack of a clear structure and norm to generate shared values.

First, the role of legislation emerged as a key driver for encouraging the involved companies to consider collaborative actions within their organizational structure, as well as motivating them to have collaboration beyond their organizational boundaries by taking financial risks. Legislation can motivate the actors to go beyond their old system and boundaries and embrace new systems or structures, as a result this may lead the new changes in the organization to fit well and be normalized in their working system.

One of the main critiques and concerns which was revealed through our study was that one business model may not be capable to meet all the participating stakeholders' values and goals, and benefit all of them equally as their core values at some point may be differing or conflicting. In this respect, the structure of a CBM can be an enabler for its performance. A CBM needs to be market-oriented and to be explicit in defining each actor's role and the proposed market. This means that a CBM should avoid having a fixed structure and instead should be flexible, that may result in developing a better response to dynamic and changing contexts, while new collaborative ideas (new CBMs within itself) will be generated. This is in line with Batonda and Perry (2003)'s state-theory which suggests that actors in a collaboration should move in a dynamic and random manner from one state to another. This is well pictured in our study, as it was stated by some actors that they foresee the forms of collaboration among them will change over time with technological advancements and changes.

Second, trust, as revealed in our study, is yet another key driver of a CBM. It is an important enabler and requirement of partners' collaboration, which can be built through open communication and knowledge sharing among partners, as has been pointed out by Larson and LaFasto (1989). Further, partners need to reach an agreement over the processes and rules of their collaboration. In addition it is important to have an assessment of the potential risks and fairness of their agreements.

Finally, an advanced and comprehensive understanding of each other's' work and core values is another key driver of a successful collaboration. Therefore, the absence of a proper understanding of each other's work can raise challenges in collaboration. For instance, actors like research institutes and charities act under different formats and strive towards different goals, and this might lead to unanticipated challenges when it comes to setting normative orientations in the CBM, such as for development of a new technology despite their common interest and support for the idea. Consequently, this calls for identifying a neutral platform in which actors can build trust, can have a dialogue, assert and share their core values and main concerns with each other. Academia with no conflicting interests with the other actors could be a good solution for that, as was identified through our study.

6.2 Contribution and Future Research Recommendations

Our study contributes with knowledge of the Swedish TWM system, by mapping the actors and activities involved with it. In doing so, a value mapping tool is being utilized which investigates the values that are missed or destroyed in the current structure, the need to create shared value among the actors, and potential challenges and opportunities to it. This knowledge suggests the potentials for development of a CBM for sustainable innovation in the context of TWM. Thus our study sheds light on how a CBM can be prescribed by utilizing a value mapping activity for sustainable innovation in TWM system. Such a CBM should be flexible in its structure and is also required to be responsive to the participators' goals and values, and be mutually beneficial. Future research can investigate the characteristics of the structure and organizational dynamics underpinning such a CBM, and how the inherent challenges can be overcome. In addition, more quantitative approaches can be taken in evaluating the positive environmental impacts and detoxification potential of higher reuse of textiles obtained through such CBMs.

Acknowledgements The research was part-funded by Re:textile, an extended regional research initiative taken by Västra Götaland Region (Region of West Sweden). The author would like to thank the funding agency for the financial support.

Appendix 1

Interview guide

Mapping main actors of TWM system in Sweden

1. From your perspective, who are the main participating actors in Swedish textile waste management system?
2. Do you recognize any gap or missed actor?
3. How do you participate in textile waste management in Sweden? What is your role and impact in terms (social, economic, environment) aspects?

Activity mapping in relation to actors interactions

1. Would you explain the core activities and values in your organization in textile waste management context?
2. Would you mention the activities you are involved with in this system?
3. Is there any collaboration among the main actors to facilitate the waste management and overcome challenges and create values from opportunities?
4. Do you have any collaboration with any other main actors within textile management system to implement your activities and reaching your values and goals?
5. Does your organization have any motivations towards collaboration with other actors?
6. In case you have any collaborative relationship with any other actor would you show that/them with the aid of an arrow?
7. How does your collaboration with other actors provide benefit to your company? (in terms of Social, economic, environment aspects)
8. What will be the requirements for collaboration with other actors?
9. What is the possible area for collaboration opportunities with other actors?
10. What are the challenges that make it difficult for actors to have collaboration?
11. What are the areas that they could gain benefit from collaboration?

Collaborative business model for sustainable innovation in TWM scenario in Sweden

1. Have you ever thought about collaborative business model for sustainable innovation as a solution provider for multi-actors business environments for the actors in order to meet their values through further collaborations?
2. What is your opinion about this business model?
3. Do you think it could be a solution for textile waste management challenges in Sweden which is multi-actors model? If yes, why? If not, why?
4. What will be the challenges and enablers of its implementation?
5. What will be the opportunities brought by this business model?
6. What is your expectation from CBM for sustainable innovation?
7. What will be your organization main focus in collaborative business model for textile waste management? What will be your role and main activities? What will be your benefit from it?

References

Afuah A (2004) Business models: a strategic management approach. McGraw-Hill/Irwin, New York, NY

Allee V (2008) Value network analysis for accelerating conversion of intangibles. J of Intellect Capital 9(1):5–24

Allee V (2011) Value networks and the true nature of collaboration. Value Net Works and Verna Allee Associates. Available at: www.valuenetworksandcollaboration.com/. Accessed 3 Mar 2013

Allwood J, Ellebæk Laursen S, Russell S, Malvido de Rodríguez C, Bocken N (2008) An approach to scenario analysis of the sustainability of an industrial sector applied to clothing and textiles in the UK. J of Cleaner Prod 16(12):1234–1246

Anderson P (2016) In: Foss NJ, Saebi T (eds) Business model innovation: the organizational dimension. OUP Oxford, Oxford

Batonda G, Perry C (2003) Approaches to relationship development processes in inter-firm networks. Eur J Mark 37(10):1457–1484

Bocken NM, Short SW, Rana P, Evans S (2013) A value mapping tool for sustainable business modelling. Corp Gov 13(5):482–497

Boulton RE, Libert BD, Samek SM (1997) Cracking the value code: how successful business are creating wealth in the new economy. Harper Collins Publisher, New York

Breuer H, Lüdeke-Freund F (2014) Normative innovation for sustainable business models in value networks. Paper presented at the 25th ISPIM conference "Innovation for Sustainable Economy and Society". Dublin, Ireland

Chesbrough H (2006) Open business models: how to thrive in the new innovation landscape. Harvard Business School Publishing, Boston

Chesbrough H (2010) Business model innovation: opportunities and barriers. Long Range Plan 43:354–363

Chesbrough H, Vanhaverbeke W, West J (2006) Open innovation: researching a new paradigm. Oxford University Press, Oxford

Christopher S, Watts W, McCormick A, Young S (2008) Building and Maintaining trust in a community-based participatory research partnership. Am J Public Health 98(8):1398–1406

Dahan NM, Doh JP, Oetzel J, Yaziji M (2010) Corporate- NGO collaboration: co-creating new business models for developing markets. Long Range Plan 43(2–3):326–342

Den Ouden E (2012) Innovation design: creating value for people organizations and society. Springer, London

Dubosson-Torbay M, Osterwalder A, Pigneur Y (2002) E-business model design, classification, and measurements. Thunderbird Int Bus Rev 44(1):5–23

Ekström K, Salomonson N (2012) Nätverk, trådar och spindlar - Samverkan for ökad återanvändning och återvinning av kläder och textil (Vetenskap för Profession:22). Högskolan i Borås, Borås

Ellen Macarthur Foundation (2014) Circular cconomy report—towards the circular economy, vol 3. Available at: https://www.ellenmacarthurfoundation.org/publications/towards-the-circular-economy-vol-3-accelerating-the-scale-up-across-global-supply-chains. Accessed 17 May 2016

Fielt E (2014) Conceptualising business models: definitions, frameworks and classifications. J Bus Models 1(1):85–105

Heikkilä M, Solaimani S, Soudunsaari A, Hakanen M, Kuivaniemi L, Suoranta M (2014) Performance estimation of networked business models: case study on a finnish ehealth service project. J Bus Models 2(1):71–88

Johnson MW, Suskewicz J (2009) How to jump-start the clean tech economy. Harvard Bus Rev 87 (11):52–60

Lambert S, Davidson R (2013) Applications of the business model in studies of enterprise success, innovation and classification: an analysis of empirical research from 1996 to 2010. Eur Manage J 31(6):668–681

Larson CE, LaFasto FMJ (1989) Team work: what must go right/what can go wrong. Sage, London

Lei DT (2000) Industry evolution and competence development: the imperatives of technological convergence. Int J Technol Manage 19(7–8):699–738

Lund M, Nielsen C (2014) The evolution of network-based business models illustrated through the case study of an entrepreneurship project. J Bus Models 2(1):105–121

Magretta J (2002) Why business models matter. Harvard Bus Rev 80(5):86–92

Naturvårdsverket (2013) Förslag till etappmål Textil och textilavfall. Stockholm: Naturvårdsverket

Nielsen C, Ahokangas P, Cöster M, Westelius A, Iveroth E, Petri C (2014) Editorial: the business model eruption and how game changing mind sets challenge existing nodes of business. J Bus Models 2(1):1–5

Osterwalder A, Pigneur Y (2002) An e-business model ontology for modeling e-business. Paper presented at the 15th bled electronic commerce conference

Osterwalder A, Pigneur Y (2010) Business model generation: a handbook for visionaries, game changers, and challengers (self-published)

Pal R (2016 (forthcoming) Extended responsibility through servitization in PSS: an exploratory study of used- clothing sector. J Fashion Mark Manage 20(4)

Palm D (2011) Improved waste management of textiles (IVL:B1976). IVL, Göteborg

Palm D, Elander M, Watson D, Kiørboe N, Salmenperä H, Dahlbo H, Moliis K, Lyng K, Valente C, Gíslason S, Tekie H, Rydberg T (2014) Towards a Nordic textile strategy (2014:450). Norden, Copenhagen

Parolini C (1999) The value net: a tool for competitive strategy. Wiley, Chichester

Porter M, Kramer M (2011) Creating shared value. Harvard Bus Rev 89(1–2): 62–77 (January–February 2011)

Rohrbeck R, Döhler M, Arnold H (2009) Creating growth with externalization of R&D results—the spin-along approach. Glob Bus Organ Excell 28(4):44–51

Rohrbeck R, Konnertz L, Knab S (2013) Collaborative business modelling for systemic and sustainability innovations. Int J Technol Manage 63(1/2):4

Ruff F (2006) Corporate foresight: integrating the future business environment into innovation and strategy. Int J Technol Manage 34(3–4):278–295

Schmidt A, Poulsen PB, Watson D, Roos S, Askham C (2016) Gaining benefits from discarded textiles: LCA of different treatment pathways. Nordic Council of Ministers, Copenhagen

Schneider S, Spieth P (2013) Business model innovation: towards an integrated future research agenda. Int J Innov Mange 17(01):1340001

Sweet P (2001) Strategic value configuration logics and the 'new' economy. A service economy revolution? Int J Serv Ind Manage 12(1):70–83

Teece D (2010) Business models, business strategy and innovation. Long Range Plan 43(2–3):172–194

Thorelli HB (1986) Networks: between markets and hierarchies. Strateg Manage J 7(1):37–51

Timmers P (1998) Business models for electronic markets. Electron Markets 8(2):3–8

Tojo N, Kogg B, Kiørboe N, Kjær B, Aalto K (2012) Prevention of textile waste: material flows of textile in three nordic countries and suggestions on policy instruments. Nordic Council of Ministers, Copenhagen

Watson D, Kiørboe N, Palm D, Tekie H, Harris S, Ekvall T, Lindhqvist T, Lyng KA (2014) EPR systems and new business models: reuse and recycling of textiles in the Nordic region. Norden, Copenhagen, vol 539

Watson D, Palm D, Brix L, Amstrup M (2016) Exports of Nordic used textiles: fate, benefits and impacts. Norden, Copenhagen, vol 558

Weill P, Vitale MR (2001) Place to space: migrating to eBusiness models. Harvard Business School Press, Boston

Wiid J, Diggines C (2009) Marketing research. Juta and Company Ltd., Claremont

WRAP (2013) Design for longevity: guidance on increasing the active life of clothing. RNF100–012

Yin RK (2014) Case study research—design and methods, 4th edn. Sage Publications, Thousand Oaks

Zamani A (2012) Developing a social business model for zero waste management systems: a case study analysis. J Environ Prot 03(11):1458–1469

Zott C, Amit R (2010) Designing your future business model: an activity system perspective. Long Range Plan 43:216–226

Zott C, Amit R, Massa L (2011) The business model: recent developments and future research. J Manage 37(4):1019–1042

Printed in the United States
By Bookmasters